PLAN® Mathematics

Test Preparation Guide

Part of the ACT® Series

ERICA DAY

COLLEEN PINTOZZI

MARY REAGAN

AMERICAN BOOK COMPANY

P. O. BOX 2638

WOODSTOCK, GEORGIA 30188-1383

TOLL FREE 1 (888) 264-5877   PHONE (770) 928-2834

TOLL FREE FAX 1 (866) 827-3240

WEB SITE: www.americanbookcompany.com

Acknowledgements

In preparing this book, we would like to acknowledge Mary Stoddard and Eric Field for their contributions in developing graphics for this book and Jennifer Reid, Breanna Bloomfield, and Clive Sombe for their contributions in editing this book. We would also like to thank our many students whose needs and questions inspired us to write this text.

Copyright © 2012
by American Book Company
P.O. Box 2638
Woodstock, GA 30188-1383

ALL RIGHTS RESERVED

The text of this publication, or any part thereof, may not be reproduced or transmitted in any form or by any means, electronic or mechanical, including photocopying, recording, storage in an information retrieval system, or otherwise, without the prior permission of the publisher.

Printed in the United States of America
03/12

PLAN is a registered trademark of ACT, Inc. American Book Company is not affiliated with ACT, Inc., and produced this book independently.

Contents

Acknowledgements   ii

Preface   xi

Diagnostic Test   1

1 About The PLAN Mathematics Test   9
- 1.1 Description of the PLAN Math Test   9
- 1.2 Content of the Test   10
- 1.3 PLAN Preparation   10
- 1.4 Taking the PLAN Mathematics Test   11
- 1.5 Types of Questions on the PLAN Mathematics Test   12

2 Numbers and Number Relations   15
- 2.1 Real Numbers   15
- 2.2 Factors   16
- 2.3 Prime Factorization   17
- 2.4 Least Common Multiple   19
- 2.5 Mixed Integer Practice   20
- 2.6 Mixed Decimal Practice   20
- 2.7 Mixed Fraction Practice   21
- 2.8 Ordering Fractions   22
- Chapter 2 Review   23
- Chapter 2 Test   24

3 Word Problems   25
- 3.1 Integer Word Problems   25
- 3.2 Fraction Word Problems   26
- 3.3 Decimal Word Problems   27
- 3.4 Estimating Rational Numbers   28
- 3.5 Exact Versus Estimated Answers   29
- 3.6 Reasonable Solutions   30
- Chapter 3 Review   31
- Chapter 3 Test   32

4 Percents   34
- 4.1 Changing Percents to Decimals and Decimals to Percents   34

Contents

4.2	Changing Percents to Fractions and Fractions to Percents	35
4.3	Percent Word Problems	36
4.4	Finding the Percent of the Total	37
4.5	Percent Increase or Decrease	38
4.6	Finding the Amount of Discount	40
4.7	Finding the Discounted Sale Price	41
	Chapter 4 Review	42
	Chapter 4 Test	42

5 Exponents and Roots — 44

5.1	Understanding Exponents	44
5.2	Multiplying Exponents with the Same Base	46
5.3	Multiplying Exponents Raised to an Exponent	46
5.4	More Multiplying Exponents	47
5.5	Negative Exponents	48
5.6	Multiplying with Negative Exponents	48
5.7	Dividing with Exponents	49
5.8	Order of Operations	50
5.9	Scientific Notation	52
5.10	Using Scientific Notation for Large Numbers	52
5.11	Using Scientific Notation for Small Numbers	53
5.12	Square Root	54
5.13	Estimating Square Roots	54
5.14	Simplifying Square Roots	55
5.15	Cube Roots	55
	Chapter 5 Review	56
	Chapter 5 Test	57

6 Ratios and Proportions — 59

6.1	Ratio Problems	59
6.2	Solving Proportions	60
6.3	Ratio and Proportion Word Problems	61
6.4	Maps and Scale Drawings	63
	Chapter 6 Review	64
	Chapter 6 Test	65

iv

Contents

7 Introduction to Graphing — 66

7.1	Graphing on a Number Line	66
7.2	Graphing Fractional Values	66
7.3	Plotting Points on a Vertical Number Line	68
7.4	Graphing Rational Numbers on a Number Line	69
7.5	Rectangle Coordinate System	71
7.6	Ordered Pairs	72
	Chapter 7 Review	74
	Chapter 7 Test	75

8 Introduction to Algebra — 76

8.1	Algebra Vocabulary	76
8.2	Substituting Numbers for Variables	77
8.3	Understanding Algebra Word Problems	78
8.4	Setting Up Algebra Word Problems	81
8.5	Changing Algebra Word Problems to Algebraic Equations	82
8.6	Substituting Numbers in Formulas	83
8.7	Properties of Addition and Multiplication	84
	Chapter 8 Review	85
	Chapter 8 Test	87

9 Equations and Inequalities — 89

9.1	Two-Step Algebra Problems	89
9.2	Two-Step Algebra Problems with Fractions	90
9.3	More Two-Step Algebra Problems with Fractions	91
9.4	Combining Like Terms	92
9.5	Solving Equations with Like Terms	92
9.6	Solving for a Variable	94
9.7	Removing Parentheses	95
9.8	Multi-Step Algebra Problems	96
9.9	Graphing Inequalities on a Number Line	98
9.10	Multi-Step Inequalities	99
9.11	Number Patterns	101
	Chapter 9 Review	102
	Chapter 9 Test	103

10 Algebra Word Problems — 104

10.1	Algebra Word Problems	104

v

10.2	Real-World Linear Equations	105
10.3	Word Problems with Formulas	107
10.4	Age Problems	108
10.5	Equivalent Forms of Equations	110
10.6	Inequality Word Problems	111
	Chapter 10 Review	112
	Chapter 10 Test	113

11 Polynomials   114

11.1	Adding and Subtracting Monomials	114
11.2	Adding Polynomials	115
11.3	Subtracting Polynomials	116
11.4	Multiplying Monomials	117
11.5	Multiplying Monomials by Polynomials	118
11.6	Removing Parentheses and Simplifying	119
11.7	Multiplying Two Binomials Using the FOIL Method	120
11.8	Simplifying Expressions with Exponents	121
	Chapter 11 Review	122
	Chapter 11 Test	123

12 Factoring   125

12.1	Finding the Greatest Common Factor of Polynomials	125
12.2	Finding the Numbers	128
12.3	More Finding the Numbers	129
12.4	Factoring Trinomials	130
12.5	More Factoring Trinomials	132
12.6	Factoring More Trinomials	133
12.7	Factoring the Difference of Two Squares	135
	Chapter 12 Review	137
	Chapter 12 Test	138

13 Solving Quadratic Equations   139

13.1	Solving Quadratic Equations	139
13.2	Solving the Difference of Two Squares	141
13.3	Solving Perfect Squares	143
13.4	Using the Quadratic Formula	144
	Chapter 13 Review	145
	Chapter 13 Test	146

Contents

14 Graphing and Writing Equations and Inequalities   147

14.1	Graphing Linear Equations	147
14.2	Graphing Horizontal and Vertical Lines	149
14.3	Finding the Distance Between Two Points	150
14.4	Finding the Midpoint of a Line Segment	151
14.5	Distance Between Points	152
14.6	Finding the Intercepts of a Line	154
14.7	Understanding Slope	155
14.8	Slope-Intercept Form of a Line	157
14.9	Verify That a Point Lies on a Line	158
14.10	Graphing a Line Knowing a Point and Slope	159
14.11	Finding the Equation of a Line Using the Slope and Y-Intercept	160
14.12	Finding the Equation of a Line Using Two Points or a Point and Slope	161
14.13	Matching Graphs of Linear Equations	162
14.14	Graphing Inequalities	163
	Chapter 14 Review	166
	Chapter 14 Test	167

15 Systems of Equations   172

15.1	Equations of Parallel Lines	172
15.2	Equations of Perpendicular Lines	173
15.3	Systems of Equations	175
15.4	Finding Common Solutions for Intersecting Lines	177
15.5	Solving Systems of Equations by Substitution	178
15.6	Solving Systems of Equations by Adding or Subtracting	180
15.7	Solving Word Problems with Systems of Equations	182
	Chapter 15 Review	185
	Chapter 15 Test	186

16 Angles   188

16.1	Types of Angles	189
16.2	Adjacent Angles	190
16.3	Vertical Angles	191
16.4	Complementary and Supplementary Angles	192
16.5	Corresponding, Alternate Interior, and Alternate Exterior Angles	193
16.6	Sum of Interior Angles of a Polygon	194
	Chapter 16 Review	195

vii

Chapter 16 Test 197

17 Triangles

198

17.1	Types of Triangles	198
17.2	Interior Angles of a Triangle	199
17.3	Exterior Angles	200
17.4	Triangle Inequality Theorem	201
17.5	Similar Triangles	202
17.6	Pythagorean Theorem	204
17.7	Finding the Missing Leg of a Right Triangle	205
17.8	Applications of the Pythagorean Theorem	206
	Chapter 17 Review	208
	Chapter 17 Test	209

18 Plane Geometry

210

18.1	Points	210
18.2	Lines, Segments, and Rays	210
18.3	Types of Polygons	211
18.4	Quadrilaterals and Their Properties	212
18.5	Perimeter	213
18.6	Area of Squares and Rectangles	214
18.7	Area of Triangles	215
18.8	Area of Trapezoids and Parallelograms	216
18.9	Circumference	217
18.10	Area of a Circle	218
18.11	Two-Step Area Problems	219
18.12	Similar Figures	221
18.13	Plane Geometry Word Problems	222
18.14	Perimeter and Area with Algebraic Expressions	223
	Chapter 18 Review	225
	Chapter 18 Test	226

19 Solid Geometry

229

19.1	Understanding Volume	229
19.2	Volume of Right Prisms	229
19.3	Volume of Spheres, Cones, Cylinders, and Pyramids	231
19.4	Two-Step Volume Problems	233
19.5	Solid Geometry Word Problems	235

Contents

	Chapter 19 Review	236
	Chapter 19 Test	238

20 Transformations — **240**

20.1	Reflections	240
20.2	Translations	243
20.3	Rotations	245
20.4	Transformation Practice	246
20.5	Dilations	247
	Chapter 20 Review	249
	Chapter 20 Test	251

21 Statistics — **252**

21.1	Range	252
21.2	Mean	253
21.3	Finding Data Missing From the Mean	253
21.4	Median	254
21.5	Mode	254
21.6	Applying Measures of Central Tendency	255
	Chapter 21 Review	256
	Chapter 21 Test	257

22 Data Interpretation — **259**

22.1	Tally Charts and Frequency Tables	259
22.2	Histograms	260
22.3	Reading Tables	261
22.4	Bar Graphs	262
22.5	Line Graphs	263
22.6	Multiple Line Graphs	264
22.7	Circle Graphs	265
22.8	Pictographs	266
22.9	Reading Venn Diagrams	267
	Chapter 22 Review	268
	Chapter 22 Test	270

23 Probability — **273**

23.1	Probability	273
23.2	More Probability	275

ix

Contents

23.3	Tree Diagrams	276
23.4	Independent and Dependent Events	278
23.5	Permutations	280
23.6	More Permutations	282
23.7	Combinations	283
23.8	More Combinations	284
	Chapter 23 Review	285
	Chapter 23 Test	287

Practice Test 1   **289**

Practice Test 2   **297**

Index   **306**

Preface

PLAN® *Mathematics Test Preparation Guide* will help you review and learn important concepts and skills related to high school mathematics. To help identify which areas are of greater challenge for you, first take the diagnostic test, and then complete the evaluation chart with your instructor in order to help you identify the chapters which require your careful attention. When you have finished your review of all of the material your teacher assigns, take the practice tests to evaluate your understanding of the material presented in this book. **The materials in this book are based on the standards in mathematics published by PLAN, Inc.**

This book contains several sections. These sections are as follows: 1) A Diagnostic Test; 2) Chapters that teach the concepts and skills for the PLAN Mathematics Test; 3) Two Practice Tests. Answers to the tests and exercises are in a separate manual.

Preface

ABOUT THE AUTHORS

Erica Day has a Bachelor of Science Degree in Mathematics and is working on a Master of Science Degree in Mathematics. She graduated with high honors from Kennesaw State University in Kennesaw, Georgia. She has also tutored all levels of mathematics, ranging from high school algebra and geometry to university-level statistics, calculus, and linear algebra. She is currently writing and editing mathematics books for American Book Company, where she has coauthored numerous books, such as *Passing the Georgia Algebra I End of Course*, *Passing the Georgia High School Graduation Test in Mathematics*, *Passing the Arizona AIMS in Mathematics*, and *Passing the New Jersey HSPA in Mathematics*, to help students pass graduation and end of course exams.

Colleen Pintozzi has taught mathematics at the elementary, middle school, junior high, senior high, and adult level for 22 years. She holds a B.S. degree from Wright State University in Dayton, Ohio and has done graduate work at Wright State University, Duke University, and the University of North Carolina at Chapel Hill. She is the author of many mathematics books including such best-sellers as *Basics Made Easy: Mathematics Review, Passing the New Alabama Graduation Exam in Mathematics, Passing the Louisiana LEAP 21 GEE, Passing the Indiana ISTEP+ GQE in Mathematics, Passing the Minnesota Basic Standards Test in Mathematics,* and *Passing the Nevada High School Proficiency Exam in Mathematics*.

Diagnostic Test

40 Minutes – 40 Questions

DIRECTIONS: Solve each problem and then choose the correct answer. Be sure to answer all the questions.

Do not linger over problems that take too much time. Solve as many as you can; then return to the others in the time you have left for this test.

You are permitted to use a calculator on this test. You may use your calculator for any problems you choose, but some of the problems may best be done without using a calculator.

Note: Unless otherwise stated, all of the following should be assumed.

1. Illustrative figures are NOT necessarily drawn to scale.

2. Geometric figures lie in a plane.

3. The word *line* indicates a straight line.

4. The word *average* indicates arithmetic mean.

---

1. Mrs. Clayson knits 10 scarves, 4 pairs of gloves, and 2 hats. If she charges $15 for scarves, $10 for gloves, and $7 for hats, how much money will she make from selling all of her products?

   A. $114
   B. $240
   C. $32
   D. $164
   E. $204

   PA

2. Nicky, a food-eating champion trains every day of the week before upcoming hotdog competitions. The number of hotdogs he ate each day was recorded as 45, 57, 42, 64, 39, 51, and 66. What was the average number of hotdogs Nicky ate per day during training?

   F. 66
   G. 52
   H. 39
   J. 57
   K. 45

   PA

3. Put the following fractions and decimals in order from least to greatest.

   $1.25, \frac{15}{16}, \frac{7}{4}, 0.09$, and $\frac{3}{2}$

   A. $1.25, \frac{3}{2}, \frac{7}{4}, 0.09, \frac{15}{16}$

   B. $0.09, \frac{15}{16}, 1.25, \frac{3}{2}, \frac{7}{4}$

   C. $1.25, \frac{7}{4}, \frac{3}{2}, 0.09, \frac{15}{16}$

   D. $0.09, 1.25, \frac{3}{2}, \frac{7}{4}, \frac{15}{16}$

   E. $\frac{7}{4}, 0.09, 1.25, \frac{3}{2}, \frac{15}{16}$

   PA

4. Solve for x: $\sqrt{(x^2 + 2x + 1)^2} = 0$

   F. $x = -1$
   G. $x = 2$
   H. infinity has many solutions
   J. F & G
   K. $x = 1$

   EA

Copyright © American Book Company

5. Solve for x: $3^x = 729$

    A. 3
    B. 4
    C. 5
    D. 6
    E. 7

EA

6. A car is moving at a rate of 1.5 miles per minute. How many miles is the car traveling per hour?

    F. 20 mph
    G. 55 mph
    H. 60 mph
    J. 90 mph
    K. 120 mph

PA

7. Solve for the missing value for the set of data if the mean is 77 and the given data values are 120, 55, 62, and 87.

    A. 64
    B. 61
    C. 77
    D. 69
    E. 59

PA

8. Combine like terms:

$$5x + 12y - 9x + 14y = 120$$

    F. $14x + 26y = 120$
    G. $40xy = 120$
    H. $-4x + 26y = 120$
    J. $x + y = 30$
    K. $4x - 26y = 120$

EA

9. What is the slope of a line containing the points $(3, 5)$ and $(8, 9)$?

    A. $\frac{4}{5}$

    B. $\frac{11}{14}$

    C. $-\frac{4}{5}$

    D. $\frac{5}{4}$

    E. $-\frac{5}{4}$

CG

10. What is the distance between the points $(1, 4)$ and $(6, 18)$? Round to the nearest hundredth.

    F. 19
    G. 12
    H. 14.86
    J. 14.87
    K. 15

CG

11. Calculate the average of the following data values: 72, 13, 15, 48, 61, 28.

    A. 40
    B. 39
    C. 28
    D. 48
    E. 39.5

PA

12. Find the measure of the angle x using the properties of parallel lines.

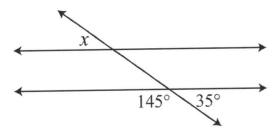

F. $145°$
G. $35°$
H. $23°$
J. $45°$
K. $120°$

PG

13. Use the Pythagorean theorem to solve for the missing side. Round to the nearest whole number.

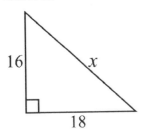

A. 24
B. 30
C. 18
D. 12
E. 17

PG

14. Identify the place value of 5 in 217,528.

F. hundred-thousands
G. ten-thousands
H. thousands
J. hundreds
K. tens

PA

15. Find the product: $(2x+3)(-x+7)$

A. $-2x^2 + 11x + 21$
B. $2x^2 - 11x - 21$
C. $2x^2 + 14x + 24$
D. $-2x^2 - 17x + 21$
E. $-2x^2 + 17x + 21$

EA

16. Compute the perimeter.

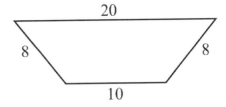

F. 30
G. 36
H. 46
J. 320
K. 160

PG

17. In the standard (x, y) coordinate plane below, 3 of the vertices of a rectangle are shown. Find the point that will be the fourth vertex of the rectangle.

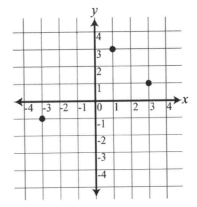

A. $(-2, -1)$
B. $(1, 1)$
C. $(-1, -3)$
D. $(0, 0)$
E. $(-4, -2)$

CG

Copyright © American Book Company

18. What is 45% of 120?

   F. 54
   G. 62
   H. 60
   J. 45
   K. 12

   PA

19. Which of the following triangles represent a Pythagorean triple?

   A.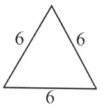
   B.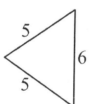
   C.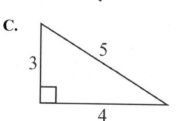
   D.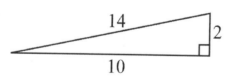
   E.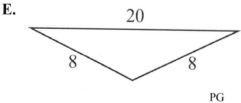

   PG

20. Compute the area of the circle below.

   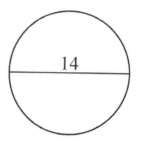

   F. 49π
   G. 7π
   H. 14π
   J. π^2
   K. 28π

   PG

21. What is the area of a right triangle with a perimeter of 30, where the base is 12 and the length of the hypotenuse is 13?

   A. 30
   B. 32.5
   C. 78
   D. 156
   E. 5

   PG

22. Find the common denominator and solve for x:

   $\frac{1}{3} + \frac{x}{18} = 2$

   F. $x = 5$
   G. $x = 10$
   H. $x = 20$
   J. $x = 30$
   K. $x = 41$

   EA

23. Write $72,100,000$ in scientific notation.

    A. 7.21×10^7

    B. 7.21×10^{-7}

    C. 72.1×10^7

    D. 72.1×10^{-7}

    E. 7.21×10^6

PA

24. Determine when the expression $\dfrac{(x-1)^2}{(x-7)}$ is undefined.

    F. $x = 1$

    G. $x = -1$

    H. $x = 7$

    J. $x = -7$

    K. $x - 0$

EA

25. Compute the area of a rectangle with a length of 12 units and a width of 14 units.

    A. 672 units^2

    B. 194 units^2

    C. 52 units^2

    D. 168 units^2

    E. 26 units^2

PG

26. Determine the slope of the line represented by $y = 2x + \frac{1}{2}$.

    F. $\frac{1}{2}$

    G. 2

    H. $\frac{3}{2}$

    J. $\frac{5}{2}$

    K. 0

CG

27. Find the midpoint of the segment between the points $(-8, 5)$ and $(2, -2)$.

    A. $(-3, \frac{3}{2})$

    B. $(3, -\frac{3}{2})$

    C. $(-3, -\frac{3}{2})$

    D. $(3, \frac{3}{2})$

    E. $(-5, \frac{7}{2})$

CG

28. If Adam buys a game for 20% off, how much will he have to pay for a game that initially costs $\$36.00$?

    F. $\$35.28$

    G. $\$7.20$

    H. $\$28.80$

    J. $\$43.20$

    K. $\$36.72$

PA

29. Find the line graph that represents the inequality $3x < 15$.

    A.

    B.

    C.

    D.

    E.

CG

Copyright © American Book Company

30. What is the measure of the missing angle?

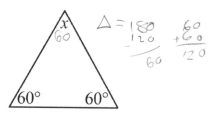

F. 30°
G. 45°
H. 60°
J. 120°
K. 180°

PG

31. What is the length of the missing side?

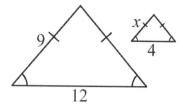

A. 4
B. 3
C. 5
D. 6
E. 9

PG

32. What is the area and perimeter of the isosceles triangle?

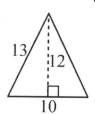

F. $A = 60, P = 30$
G. $A = 60, P = 60$
H. $A = 30, P = 60$
J. $A = 60, P = 36$
K. $A = 36, P = 30$

PG

33. Compute the perimeter of the simple composite geometric figure with unknown side lengths.

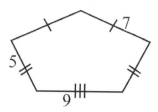

A. 21
B. 24
C. 33
D. 23
E. 19

PG

34. Calculate the average of 7 numbers whose sum is 959.

F. 137
G. 49
H. 142
J. 154
K. 136

PA

35. Which of the following fractions is equivalent to $\frac{6}{7}$?

A. $\frac{75}{100}$

B. $\frac{54}{63}$

C. $\frac{25}{50}$

D. $\frac{4}{3}$

E. $\frac{7}{6}$

PA

36. If there are 7 red marbles, 4 blue marbles, and 6 green marbles, what are the chances that a person will select a green marble if they blindly drew from the bag of marbles?

   F. $\frac{6}{17}$
   G. $\frac{4}{17}$
   H. $\frac{7}{17}$
   J. $\frac{2}{3}$
   K. $\frac{1}{2}$

PA

37. What is the cube root of 343?

   A. 5
   B. 6
   C. 7
   D. 12
   E. 13

PA

38. Solve the linear inequality:

   $-7x < 2401 - 49$

   F. $x < 49$
   G. $x < -336$
   H. $x > 49$
   J. $x < 343$
   K. $x > -336$

EA

39. Solve for x: $x^2 - \frac{7}{4}x - \frac{1}{2} = 0$.

   A. $x = \frac{1}{4}, 2$
   B. $x = -\frac{1}{4}, 2$
   C. $x = \frac{1}{2}, \frac{1}{4}$
   D. $x = -\frac{1}{2}, -\frac{1}{4}$
   E. $x = \frac{1}{2}, -\frac{1}{4}$

EA

40. Locate $(-3, -2)$ on the coordinate plane.

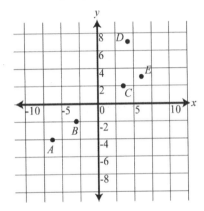

   F. Point A
   G. Point B
   H. Point C
   J. Point D
   K. Point E

CG

Evaluation Chart for the Diagnostic Mathematics Test

Directions: On the following chart, circle the question numbers that you answered incorrectly. Then turn to the appropriate topics (listed by chapters), read the explanations, and complete the exercises. Review the other chapters as needed. Finally, complete PLAN Mathematics Test Preparation Guide Practice Tests for further review.

		Questions	Pages
Chapter 2:	Numbers and Number Relations	3, 14, 35	15–24
Chapter 3:	Word Problems	1	25–33
Chapter 4:	Percents	18, 28	34–43
Chapter 5:	Exponents and Roots	5, 23, 37	44–58
Chapter 6:	Ratios and Proportions	6	59–65
Chapter 7:	Introduction to Graphing	40	66–75
Chapter 8:	Introduction to Algebra		76–88
Chapter 9:	Equations and Inequalities	8, 22, 29, 38	89–103
Chapter 10:	Algebra Word Problems		104–113
Chapter 11:	Polynomials	15	114–124
Chapter 12:	Factoring	24	125–138
Chapter 13:	Solving Quadratic Equations	4, 39	139–146
Chapter 14:	Graphing and Writing Equations and Inequalities	9, 10, 26, 27	147–171
Chapter 15:	Systems of Equations		172–187
Chapter 16:	Angles	12	188–197
Chapter 17:	Triangles	13, 19, 30, 31	198–209
Chapter 18:	Plane Geometry	16, 20, 21, 25, 32, 33	210–228
Chapter 19:	Solid Geometry		229–239
Chapter 20:	Transformations	17	240–251
Chapter 21:	Statistics	2, 7, 11, 34	252–258
Chapter 22:	Data interpretation		259–272
Chapter 23:	Probability	36	273–288

8   Copyright © American Book Company

Chapter 1
About The PLAN Mathematics Test

1.1 Description of the PLAN Math Test

The PLAN Mathematics Test is a 40-item, 40-minute test. The following four categories represent content areas of the PLAN Mathematics Test commonly taught by the end of grade 10 that are important to success in entry-level college mathematics courses:

- **Pre-Algebra**

- **Elementary Algebra**

- **Coordinate Geometry**

- **Plane Geometry**

You will receive a score for all 40 questions and two subscores: a Pre-Algebra/Elementary Algebra subscore based on 22 questions, and a Coordinate Geometry/Plane Geometry subscore based on 18 questions.

The PLAN Mathematics Test measures the student's level of mathematical achievement. It emphasizes quantitative reasoning rather than memorization of formulas or computational skills. In particular, it emphasizes the ability to solve practical quantitative problems that require skills encountered in many first- and second-year high school courses (pre-algebra, first- and second-year algebra, and plane geometry). While some material from second-year courses is included on the test, most items, including the geometry items, emphasize content presented before the second year of high school.

The items in the PLAN Mathematics Test cover four cognitive levels: **knowledge and skills**, , **understanding concepts**, and **integrating conceptual understanding**. **Knowledge and skills items** require the student to use one or more facts, definitions, formulas, or procedures to solve straightforward problems set in real-world situations. **Direct application items** are word problems from everyday life that need mathematics to solve. **Understanding concepts items** test the student's depth of understanding major concepts by requiring reasoning from a concept to reach an inference or a conclusion. **Integrating conceptual understanding items** test the student's ability to achieve an integrated understanding of two or more major concepts to solve non-routine problems.

Copyright © American Book Company

9

Chapter 1   About The PLAN Mathematics Test

1.2   Content of the Test

Items are classified according to the four content areas mentioned in the previous section. These categories and the approximate proportion of the test devoted to each are given in the following table.

ACT Mathematics Test 60 items, 60 minutes		
Content Area	**Proportion of Test**	**Number of Items**
Pre-Algebra	0.35	14
Elementary Algebra	0.20	8
Coordinate Geometry	0.18	7
Plane Geometry	0.27	11
Total	**1.00**	**40**

1. **Pre-Algebra.** Items in this content area are based on operations using whole numbers, decimals, fractions, and integers; place value; square roots and approximations; the concept of exponents; scientific notation; factors; ratio, proportion, and percent; linear equations in one variable; absolute value and ordering numbers by value; elementary counting techniques and simple probability; data collection, representation, and interpretation; and understanding simple descriptive statistics.

2. **Elementary Algebra.** Items in this content area are based on properties of exponents and square roots, evaluation of algebraic expressions through substitution, simplification of algebraic expressions; addition, subtraction, and multiplication of polynomials; and solving quadratic equations by factoring.

3. **Coordinate Geometry.** Items is this content area are based on graphing and the relations between equations and graphs, including points and lines; graphing inequalities; slope; parallel and perpendicular lines; distance; and midpoints.

4. **Plane Geometry**. Items in this content area are based on the properties and relations of plane figures, including angles and relations among perpendicular and parallel lines; properties of circles, triangles, rectangles, parallelograms, and trapezoids; transformations; and volume.

Students are permitted but not required to use calculators when taking this test. If they do so, they should use the calculator they are most familiar with. All of the problems can be solved without a calculator.

1.3   PLAN Preparation

The PLAN measures your overall learning, so if you have paid attention in school, you should do well. It would be difficult to "cram" for an exam as comprehensive as this. However, you can study wisely by using an PLAN-specific guide and practice answering questions of the type that will be asked on the PLAN.

Believe in yourself! Attitude plays a big part in how well you do in anything. Keep your thoughts positive. Tell yourself you will do well on the exam.

10   Copyright © American Book Company

Be prepared. Get a good night's sleep the day before your exam. Eat a well-balanced meal, one that contains plenty of proteins and carbohydrates, prior to your exam.

Arrive early. Allow yourself at least 15–20 minutes to find your room and get settled. Then you can relax before the exam, and you won't feel rushed.

Practice relaxation techniques. Many students become stressed, and they begin to worry too much about the exam. They may perspire heavily, experience upset stomach, or have shortness of breath. If you feel any of these symptoms, talk to a close friend or see your counselor. They will suggest ways to deal with your test anxiety.

1.4   Taking the PLAN Mathematics Test

Read the instructions on the PLAN test booklet carefully. To ensure that you will understand the instructions, you can read them at www.actstudent.org prior to taking the test.

Once you are told that you may open the booklet, read the directions thoroughly for the test before beginning to mark your answers.

Read each question carefully enough so that you know what you're trying to find, and use your best approach for answering the questions.

Use your calculator wisely and sparingly. Remember, some problems can be solved without using a calculator. In fact, some of the problems are best done without a calculator. Use good judgment to decide whether or not you need to use your calculator.

Solve the problem. For working out the solutions to the problems, you may usually do scratch work in the space provided in the test booklet, or you will be given scratch paper to use.

Answer every question on the exam. Your score is based on the number of questions you answer correctly. There is no penalty for guessing and no penalty for wrong answers, but every spot left blank is an automatic zero. A guess has a 20% chance of being correct, whereas a blank has no chance of being correct. So, if you are unsure about an answer, take an educated guess. Eliminate choices that are definitely wrong, and then choose from the remaining answers.

Use your answer key correctly. Make sure the number on your question matches the number on your answer sheet. If you need to change an answer, erase it completely. Use a number two pencil. The computerized scanner may skip over answers that are too light so make sure the answers are dark.

Check your answers. Make sure your answers make sense. If you finish the test before time is called, review your exam to make sure you have chosen the best responses. Change answers only if you are sure they are wrong.

Be sure to pace yourself. Since you will have a limited amount of time, be careful not to spend too much time on one problem, leaving no time to complete the rest of the test. Keep an eye on the clock to make sure that you are working at a pace that will allow you to finish the test in the 60 minutes given. Listen for announcement of five minutes remaining on the test.

When time is called for the test, put your pencil down. If you continue to write or erase, you run the risk of being dismissed and your test being disqualified from scoring.

Copyright © American Book Company

Chapter 1   About The PLAN Mathematics Test

1.5   Types of Questions on the PLAN Mathematics Test

The content of the PLAN Mathematics Test will vary. The questions vary in difficulty and complexity. The type of questions include basic math problems, basic math problems in settings, very challenging problems, and question sets.

Basic Math Problems

Basic math problems are simple and straightforward. They usually have very few words, no extra information, and a numeric solution. Question 1 is an example of a basic pre-algebra math problem.

1. What is 4% more than $1,250$?

    A. $1,250$
    B. $1,200$
    C. $1,300$
    D. $1,365$
    E. $1,380$

This problem has few words, a direct question is asked, and its answer is numeric. The solution is simple: change 4% to a decimal and multiply by $1,250$ to get $(0.04)(1,250) = 50$ Add this number to the number in the question. (**C**).

Basic Math Problems in Settings

Basic math problems in settings are often called **word problems** or **story problems**. They describe situations from everyday life where you need to apply mathematics to find an answer to a question. Question 2 is an example of a basic elementary algebra problem in a setting.

2. Alicia bought 10 pounds of potatoes for $3.99. She also purchased 5 pounds of onions. The total for Alicia's groceries is $6.98, not including taxes. How much was Alicia charged for the onions, not including taxes?

    F. $2.99
    G. $2.89
    H. $3.19
    J. $3.99
    K. $2.88

In this problem, you know the total amount of the two items and the amount of one item. Use subtraction to find the price of the other item: $6.98 - \$3.99 = \$2.99$. (**F**)

12   Copyright © American Book Company

1.5 Types of Questions on the PLAN Mathematics Test

Very Challenging Problems

These problems are not like the problems you usually see. They test your ability to reason and appear in all different forms. These challenging problems will show up in all four categories of the test. Question 3 is an example of a very challenging intermediate algebra problem.

3. If the integer $3 \times 4^x = 192$, then $x = ?$

   A. 6
   B. 5
   C. 2
   D. 3 ←
   E. 16

In this problem, you're asked to find the value of x. Divide both sides by 3: $4^x = 64$ Find how many times you must multiply 4 to equal 64: $4 \times 4 \times 4 = 64$. Therefore, $x = 3$. **(D)**.

Question Sets

The test contains sets of questions that all relate to the same information. There are usually two question sets on the test, with two or four questions on each set. Questions 4–5 show a group of questions that use the same information. This is called a question set.

Use the following information to answer questions 4 and 5.

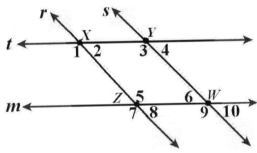

4. Which two angles add up to $180°$?

   F. angles 1 and X
   G. angles 1 and 8
   H. angles X and Y
   J. angles Z and 8
   K. angles Z and 2

5. If angle Y is equal to $130°$, what is the measure of angle 4?

   A. $40°$
   B. $65°$
   C. $230°$
   D. $45°$
   E. $50°$

Copyright © American Book Company

Chapter 1   About The PLAN Mathematics Test

Question 4 requires you to two angles that add up to 180°. You would need to know that a straight line measures 180° and then find two angles from the list that equal a straight line when next to each other.(**G**).

Question 5 asks you to find the measure of angle 4. You can see by the drawing that angle Y and angle 4 put next to each other makes a straight angle. Questions 5 also requires you to know that a straight angle measures 180°. If you know the measure of $Y = 130°$, then you can subtract to find the measure of angle 4: $180° - 130° = 50°$ (**C**).

The diagnostic test at the beginning and two practice tests at the end of the book are simulated PLAN tests. They are the same length and contain questions comparable to those you will see on the PLAN. Review your scores on the diagnostic test and the practice tests with your teacher or tutor to determine if there are skill areas you need to focus on before taking the PLAN.

For practice with other sections of the PLAN, refer to these titles from American Book Company:

PLAN English Test Preparation Guide

PLAN Reading Test Preparation Guide

PLAN Science Test Preparation Guide

Chapter 2
Numbers and Number Relations

2.1 Real Numbers

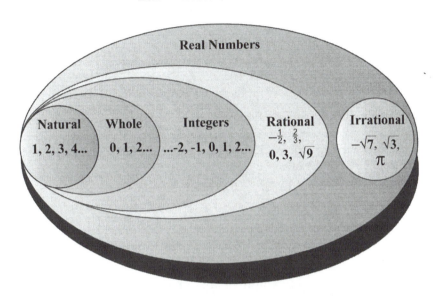

Real numbers include all positive and negative numbers and zero. Included in the set of real numbers are positive and negative fractions, decimals, and rational and irrational numbers.

Use the diagram above and your calculator to answer the following questions.

1. Using your calculator, find the square root of 5. Does it repeat? Does it end? Is it a rational or an irrational number?
2. Find $\sqrt{49}$. Is it rational or irrational? Is it an integer?
3. Is an integer a rational number?
4. Is an integer a whole number?
5. Is $\frac{7}{4}$ a real number? Is it rational or irrational?
6. Is zero a natural number?

Identify the following numbers as rational (R) or irrational (I).

7. 6π
8. $\sqrt{25}$
9. $\frac{6}{11}$
10. $\sqrt{12}$
11. 0.013

12. -4
13. $\frac{1}{9}$
14. $\pi + 2$
15. $-\frac{13}{15}$
16. 6.1011

17. $\sqrt{29}$
18. $\frac{4}{15}$
19. $-\sqrt{64}$
20. 62.22

21. $\frac{11}{12}$ R
22. $-\sqrt{42}$ I
23. -5.7 I
24. π I

Copyright © American Book Company

15

Chapter 2   Numbers and Number Relations

25. What is the difference between an integer and a whole number? + and – numbers

26. Are natural numbers whole numbers? No, zero's included

27. What is the difference between irrational numbers and integers?

28. Are irrational numbers real numbers?

2.2   Factors

Factor: A number that is multiplied to get a product.

Example 1:   2 is a factor of 4 because $2 \times 2 = 4$

factors   product

Example 2:   List all factors for the number 16.

Step 1:   How many ways can we multiply 2 whole numbers to gain the product 16? Review the chart below.

Factor		Factor		Product
1	×	16	=	16
2	×	8	=	16
4	×	4	=	16
8	×	2	=	16

Step 2:   Without duplicating the numbers listed in the chart above, we have 1, 2, 4, 8, 16. These numbers are factors of 16.
The factors of 16 are 1, 2, 4, 8, and 16.

Example 3:   List all factors for the number 24.

Factor		Factor		Product
1	×	24	=	24
2	×	12	=	24
3	×	8	=	24
4	×	6	=	24
6	×	4	=	24
8	×	3	=	24
12	×	2	=	24
24	×	1	=	24

Answer: 1, 2, 3, 4, 6, 8, 12, and 24

List all the factors for the numbers below.

1. 6

2. 15

3. 11 prime #

4. 27

5. 45

6. 99

7. 100

8. 10

9. 13

10. 14

11. 4

12. 32

2.3 Prime Factorization

Prime factorization is the process of factoring a number into prime numbers. A prime number, also called a prime, is a number that can only be divided by itself and 1. There are two main ways of finding the primes of a number: dividing and splitting.

Example 4: Find the primes of 66 by division.

Step 1: To find the primes by division, you must only divide 66 by prime numbers until you can only divide by one.

$66 \div 2 = 33, 33 \div 3 = 11, 11 \div 11 = 1, 11 \div 1 = 11$ (1 is not prime)

Step 2: All the prime numbers used as divisors make up the prime factorization of 6.
$66 = 2 \times 3 \times 11$

Check: To check, multiply the prime numbers together, and you should get the original value.

Example 5: Find the primes of 66 and 120 using the splitting method.

Step 1: In this method, the number must be split by any two factors until all of the factors are prime.

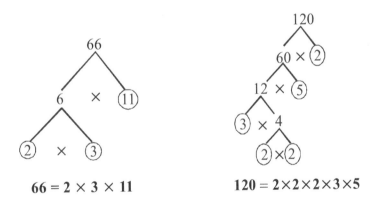

$66 = 2 \times 3 \times 11$ $120 = 2 \times 2 \times 2 \times 3 \times 5$

Step 2: All the prime numbers used in splitting 66 make up the prime factorization of 66. All the prime numbers used in splitting 120 make up the prime factorization of 120.

Hint: The factors found during prime factorization should always be prime, should always multiply together to get the correct answer, and should always be listed from least to greatest.

Chapter 2   Numbers and Number Relations

Find the prime factorization of each number using the division method. The first one has been done for you.

1. $10 \div 2 = 5 \div 5 = 1$
   $10 = 2 \times 5$

2. 14

3. 55

4. 110

5. 126

6. 142

7. 8

8. 21

9. 32

10. 36

11. 51

12. 84

13. 125

14. 48

15. 77

16. 65

17. 200

18. 413

Find the prime factorization of each number using the splitting method. The first one has been done for you.

19.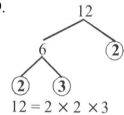
    $12 = 2 \times 2 \times 3$

20. 24

21. 45

22. 120

23. 52

24. 91

25. 18

26. 67

27. 20

28. 15

29. 35

30. 122

2.4   Least Common Multiple

To find the **least common multiple (LCM)**, of two numbers, first list the multiples of each number. The multiples of a number are 1 times the number, 2 times the number, 3 times the number, and so on.

The multiples of 6 are: 6, 12, 18, 24, 30...

The multiples of 10 are: 10, 20, 30, 40, 50...

What is the smallest multiple they both have in common?   30

30 is the **least** (smallest number) **common multiple** of 6 and 10.

Find the least common multiple (LCM) of each pair of numbers below.

	Pairs	Multiples	LCM		Pairs	Multiples	LCM
1.	6	6, 12, 18, 24, 30	30	10.	6		
	15	15, 30			7		
2.	12			11.	4		
	16				18		
3.	18			12.	7		
	36				5		
4.	7			13.	30		
	3				45		
5.	12			14.	3		
	8				8		
6.	6			15.	12		
	8				9		
7.	4			16.	5		
	14				45		
8.	9			17.	3		
	6				5		
9.	2			18.	4		
	15				22		

Copyright © American Book Company

Chapter 2   Numbers and Number Relations

2.5   Mixed Integer Practice

Simplify the following expressions.

1. $(-3) + 14 =$

2. $(-4) + (-12) =$

3. $(-6) \times 3 =$

4. $(-28) \div 7 =$

5. $(-2) - 1 =$

6. $(-1) \times (-5) =$

7. $9 + (-3) =$

8. $5 + (-18) =$

9. $(-8) - (-36) =$

10. $4 + (-1) =$

11. $\dfrac{(-18)}{(-2)} =$

12. $(-6) + 4 =$

13. $(-8) - 7 =$

14. $(-21) \div (-7) =$

15. $(-4)(-3) =$

16. $(-9) + (-28) =$

17. $(-6) - 1 =$

18. $12(-6) =$

19. $(-3) \cdot (-3) =$

20. $10 + (-4) =$

21. $\dfrac{-24}{3} =$

2.6   Mixed Decimal Practice

Simplify the following expressions.

1. $7.8 + (-3.02) + 1.2$

2. $7.4 \div (-0.001)$

3. $9.3 + 4.17 + 0.175$

4. $37.091 - 30.07$

5. $11.5 + 9.5 + 0.011$

6. $191.06 \div (-2)$

7. 12.2×4.9

8. $\$3.25 + \$8.99$

9. $\$5.54 \times (-0.07)$

10. $\$0.31 + \$4.78$

11. 1.9×0.17

12. $-10.1 + 6.05 + 9.025$

13. 6.37×2.15

14. $82.14 \div 5$

15. $1.7 + 6.32 + 6.935$

16. $\$7.15 + \$7.15$

17. $\$25.49 - \$0.78$

18. $(-12.3) \times (-1.9)$

19. $136.209 - 14.1$

20. 0.057×1.7

21. $19.7 - 16.2$

22. $8.25 - 7.001$

23. $-141.9 - 17.865$

24. 6.5×7.83

25. $47.93 \div 2$

26. -21.07×1.5

27. $13.78 \div 0.08$

28. $9.35 + 0.001 + (-1.2)$

29. $6.2 - 4.97$

30. 37.2×4.6

31. $\$7.82 \div 0.01$

32. -0.074×18

33. $45.6 \div 0.005$

34. 4.7×0.003

35. $45.6 \times (-0.113)$

36. $32.1 \div 0.008$

20   Copyright © American Book Company

2.7 Mixed Fraction Practice

Simplify the following expressions.

1. $4\frac{1}{2} \times \left(-2\frac{2}{3}\right)$

2. $8\frac{1}{4} \div \frac{3}{8}$

3. $\frac{4}{5} \times 6\frac{1}{8}$

4. $-2\frac{5}{12} + 1\frac{4}{9}$

5. $7 - 2\frac{3}{5}$

6. $3\frac{1}{9} \times 2\frac{2}{7}$

7. $(-3) \times 2\frac{1}{4}$

8. $28 - 21\frac{5}{8}$

9. $5\frac{1}{8} - 3\frac{3}{4}$

10. $1\frac{1}{8} \times \left(-1\frac{3}{4}\right)$

11. $-6 \div \frac{3}{4}$

12. $2\frac{3}{4} + \frac{7}{8}$

13. $\frac{4}{7} \times \frac{6}{13} =$

14. $7\frac{5}{9} \div 7\frac{4}{9}$

15. $\left(-4\frac{7}{8}\right) + 3\frac{1}{9}$

16. $5\frac{3}{5} \times 3\frac{1}{7}$

17. $4\frac{11}{12} - 3\frac{1}{3}$

18. $6\frac{2}{3} \times \frac{9}{10}$

19. $2\frac{1}{2} \div \frac{5}{8}$

20. $7\frac{1}{2} \div \left(-4\frac{1}{8}\right)$

21. $3\frac{2}{5} \times 1\frac{1}{4}$

22. $5\frac{1}{3} \div 2\frac{2}{9}$

23. $\frac{3}{4} \div 2$

24. $2\frac{3}{4} + 5\frac{1}{8}$

25. $4 \div 1\frac{1}{7}$

26. $15\frac{3}{10} - \left(-7\frac{1}{2}\right)$

27. $2\frac{1}{5} \div 3\frac{2}{5}$

28. $1\frac{5}{8} + \frac{3}{4}$

29. $8\frac{2}{7} - 2\frac{1}{2}$

30. $-5\frac{5}{9} \div 1\frac{1}{3}$

31. $24 - 12\frac{1}{2}$

32. $-1\frac{1}{2} \div \frac{3}{4}$

33. $7\frac{4}{6} - 4\frac{2}{3}$

34. $13 - 11\frac{2}{3}$

35. $8\frac{5}{6} + \left(-3\frac{7}{10}\right)$

36. $3\frac{1}{5} - 1\frac{2}{15}$

Copyright © American Book Company

21

Chapter 2   Numbers and Number Relations

2.8   Ordering Fractions

If the four fractions are given $\frac{3}{10}, \frac{5}{6}, \frac{1}{2}$, and $\frac{9}{10}$, the greatest, $\frac{9}{10}$, and least, $\frac{3}{10}$, are easy to find. If the four fractions given are $\frac{3}{5}, \frac{5}{6}, \frac{1}{2}$, and $\frac{11}{15}$, you probably need to find a common denominator first to figure out which fraction is the greatest or the least.

$$\frac{3}{5} = \frac{18}{30}$$

$$\frac{5}{6} = \frac{25}{30}$$

$$\frac{1}{2} = \frac{15}{30}$$

$$\frac{11}{15} = \frac{22}{30}$$

Ask yourself what number you can divide by 5, 6, 2, and 15 without a remainder. $5\overline{)?} \quad 6\overline{)?} \quad 2\overline{)?} \quad 15\overline{)?}$. The number 30 is the smallest number that can be divided by 5, 6, 2, and 15. Once equivalent fractions are found with the same denominator, you can look at the top number to tell which fraction is greatest, $\frac{5}{6}$ or least, $\frac{1}{2}$.

Circle the greatest fraction and underline the least fraction in each group below.

1. $\frac{1}{4} \quad \frac{3}{8} \quad \frac{1}{12} \quad \frac{5}{6}$

2. $\frac{3}{5} \quad \frac{7}{10} \quad \frac{11}{20} \quad \frac{3}{4}$

3. $\frac{2}{3} \quad \frac{1}{6} \quad \frac{4}{9} \quad \frac{1}{2}$

4. $\frac{67}{100} \quad \frac{13}{25} \quad \frac{3}{10} \quad \frac{4}{5}$

5. $\frac{7}{9} \quad \frac{5}{12} \quad \frac{3}{4} \quad \frac{5}{8}$

6. $\frac{5}{18} \quad \frac{1}{4} \quad \frac{5}{9} \quad \frac{1}{2}$

7. $\frac{5}{6} \quad \frac{1}{2} \quad \frac{3}{4} \quad \frac{7}{12}$

8. $\frac{13}{36} \quad \frac{3}{4} \quad \frac{3}{18} \quad \frac{7}{9}$

9. $\frac{6}{7} \quad \frac{2}{3} \quad \frac{10}{21} \quad \frac{1}{2}$

10. $\frac{5}{16} \quad \frac{7}{8} \quad \frac{1}{4} \quad \frac{1}{2}$

Find common denominators to determine which fraction is between the two given fractions. Circle the answer.

11. Which fraction is between $\frac{1}{3}$ and $\frac{5}{8}$? $\frac{2}{3}, \frac{3}{4}, \frac{5}{6}$, or $\frac{3}{8}$

12. Which fraction is between $\frac{1}{2}$ and $\frac{5}{8}$? $\frac{1}{3}, \frac{3}{4}, \frac{3}{8}$ or $\frac{9}{16}$

13. Which fraction is between $\frac{3}{7}$ and $\frac{7}{8}$? $\frac{1}{4}, \frac{3}{4}, \frac{5}{14}$ or $\frac{25}{28}$

Chapter 2 Review

Simplify the following expressions.

1. $\frac{4}{8} + \left(-\frac{1}{8}\right) =$

2. $2 - (-3) =$

3. $27 \div (-9) =$

4. $7 - 2\frac{1}{5} =$

5. $(-2) \cdot (10) =$

6. $-7 + (-1) =$

7. $1\frac{1}{5} \times \left(-2\frac{11}{12}\right) =$

8. $(-16) \div (-4) =$

9. $\left(-8\frac{3}{4}\right) + 1\frac{2}{9} =$

10. $2\frac{1}{2} \times 4\frac{3}{5} =$

11. $14.37 + 2.091 + 9.1 =$

12. $5 + (-5) =$

13. $\left(-6\frac{1}{2}\right) + \left(-\frac{5}{6}\right) =$

14. $2\frac{1}{3} - \frac{1}{6} =$

15. $39.7 - 24.97 =$

16. $15.06 \times 0.31 =$

17. $10\frac{1}{2} - \left(-7\frac{2}{3}\right) =$

18. $(2)(-2) =$

19. $3\frac{3}{4} - \frac{1}{4} =$

20. $12.242 - 8.73 =$

21. $\frac{2}{9} \times 3\frac{1}{6} =$

22. $17.241 + 21.3 + 0.004 =$

23. $41.2 \times 1.09 =$

24. $44.96 \div 1.32 =$

25. $-\frac{5}{9} \div 1\frac{1}{3} =$

26. $-8 - 3 =$

27. $(4)(-2) =$

28. $1\frac{1}{2} \div \frac{3}{4} =$

29. $\left(-11\frac{2}{3}\right) \div \left(-4\frac{1}{6}\right) =$

30. $3 - (-3) =$

31. $2.095 - 0.133 =$

32. $9 + 3 =$

33. $-4.167 \div 3 =$

34. $(-3) \times (-5) =$

35. $\frac{2}{5} + \frac{3}{5} =$

36. $12 - 15 =$

37. What is the difference between whole numbers and rational numbers?

38. Are natural numbers real numbers?

39. Is $\frac{7}{9}$ a real number? What best describes $\frac{7}{9}$?

40. What is the difference between natural numbers and whole numbers?

Label the following numbers as C for composite or P for prime. List all factors.

41. 15 _____

42. 16 _____

43. 19 _____

44. 20 _____

Find the least common multiple for the following pairs of numbers.

45. 8 and 12   46. 5 and 9   47. 4 and 10   48. 6 and 8

Circle the greatest fraction and underline the least fraction in each group below.

49. $\frac{1}{3}$ $\frac{3}{8}$ $\frac{1}{2}$ $\frac{3}{4}$

50. $\frac{2}{9}$ $\frac{7}{18}$ $\frac{1}{2}$ $\frac{1}{18}$

Copyright © American Book Company

Chapter 2   Numbers and Number Relations

Chapter 2 Test

1. Which best describes 3π?

   A an integer
   B a natural number
   C an irrational number
   D a whole number
   E a rational number

2. Simplify the following expression: $\frac{4}{5}+\left(-\frac{2}{9}\right)$

   F. $-\frac{1}{2}$

   G. $\frac{2}{14}$

   H. $-\frac{3}{2}$

   J. $\frac{26}{45}$

   K. $\frac{8}{14}$

3. Which best describes $-\frac{4}{13}$?

   A an integer
   B a natural number
   C an irrational number
   D a whole number
   E a rational number

4. Simplify the following expression:
   $(-62) \div (-2)$

   F. -31
   G. -124
   H. 124
   J. 31
   K. 62

5. Simplify the following expression: $\frac{4}{5} \times \frac{9}{10}$

   A $\frac{18}{25}$

   B $\frac{36}{50}$

   C $\frac{13}{15}$

   D $\frac{72}{10}$

   E $\frac{36}{5}$

6. Simplify the following expression:
   5.7×0.12

   F. 68.4
   G. 5.82
   H. 0.684
   J. 58.2
   K. 6.84

7. Simplify the following expression: $-6\frac{1}{3} \div \frac{3}{5}$

   A $-\frac{85}{9}$

   B $-10\frac{5}{9}$

   C $10\frac{5}{9}$

   D $-9\frac{4}{9}$

   E $\frac{95}{9}$

8. Which number is a factor of the number 32?

   F. 14
   G. 5
   H. 7
   J. 16
   K. 3

9. What is the prime factorization for the number 18?

   A. $2 \times 3 \times 3$
   B. 9×2
   C. 6×3
   D. $6 \times 3 \times 1$
   E. $2 \times 2 \times 3 \times 3$

10. What is the LCM for the numbers 5 and 6?

   F. 30
   G. 25
   H. 60
   J. 80
   K. 36

24   Copyright © American Book Company

Chapter 3
Word Problems

3.1   Integer Word Problems

Carefully read the word problems below and answer.

1. Carrie Ann has a penny bank with 812 pennies in it. She gave her little sister 10 pennies to start her own penny bank savings. She gave 68 pennies to her brother in exchange for taking her turn at doing the dishes. How many pennies does Carrie Ann have left in her bank?

2. Henri owns a cupcake shop that also takes special orders. Henri starts the day baking and frosting 120 dozen cupcakes before the store opens. On Wednesday, he received a special order for 20 dozen assorted cupcakes to be delivered that same afternoon. He also sold 98 dozen cupcakes that day. How many dozen cupcakes does Henri have left at the end of the day?

3. The temperature on a January Monday morning started at $-23°$. By noon, the temperature went up $17°$, and up an additional $12°$ by 3:00 PM. At 8:00 PM, the temperature fell $19°$. What is the temperature at 8:00 PM?

4. Sharikka and her friends were playing a game that included a pile of cards telling them to move forward or backwards a certain number of spaces. Sharikka sorted the cards by the amount written on each card. She found six cards that said "move -3 spaces." If one player received all six cards that said "move -3 spaces," how many spaces would that player move?

5. Ira and Mario are racing each other through a bonus round in a math competition. The problem they are given is: $-81 \div 9 =$ ___ . Ira rings the bell first and gives an answer of 8. Mario rings in and gives an answer of -9. Who is correct?

Copyright © American Book Company

25

Chapter 3 Word Problems

3.2 Fraction Word Problems

Solve and reduce answers to lowest terms.

1. Sara works for a movie theater and sells candy by the pound. Her first customer bought $2\frac{1}{8}$ pounds of candy, the second bought $\frac{4}{5}$ of a pound, and the third bought $\frac{1}{2}$ pound. How many pounds did she sell to the first three customers?

2. Beth has a bread machine that makes a loaf of bread that weighs $2\frac{1}{2}$ pounds. If she makes a loaf of bread for each of her three sisters, how many pounds of bread will she make?

3. A farmer hauled in 81 bales of hay. Each of his cows ate $\frac{3}{4}$ bales. How many cows did the farmer feed?

4. Juan was competing in a 1000 meter race. He had to pull out of the race after running $\frac{7}{8}$ of it. How many meters did he run?

5. Tad needs to measure where the free throw line should be in front of his basketball goal. He knows his feet are $1\frac{1}{4}$ feet long and the free-throw line should be 15 feet from the backboard. How many toe-to-toe steps does Tad need to take to mark off 15 feet?

6. A chemical plant takes in $6\frac{1}{3}$ million gallons of water from a local river and discharges $2\frac{7}{12}$ million back into the river. How much water does not go back into the river?

7. In January, Jeff filled his car with $12\frac{1}{8}$ gallons of gas the first week, $10\frac{3}{4}$ gallons the second week, $11\frac{1}{2}$ gallons the third week, and $10\frac{1}{2}$ gallons the fourth week. How many gallons of gas did he buy in January?

8. Li Tun makes sandwiches for his family. He has $9\frac{3}{4}$ ounces of sandwich meat. If he divides the meat equally to make $4\frac{1}{2}$ sandwiches, how much meat will each sandwich have?

9. The company water cooler started with $5\frac{3}{8}$ gallons of water. Employees drank $2\frac{1}{4}$ gallons. How many gallons were left in the cooler?

10. Rita bought $\frac{1}{3}$ pound hamburger patties for her family reunion picnic. She bought 40 patties. How many pounds of hamburger did she buy?

26 Copyright © American Book Company

3.3 Decimal Word Problems

1. Micah can have his oil changed in his car for $24.99, or he can buy the oil and filter and change it himself for $11.43. How much would he save by changing the oil himself?

2. Megan bought 6 boxes of cookies for $2.25 each. How much did she spend?

3. Will subscribes to a monthly auto magazine. His one year subscription costs $19.97. If he pays for the subscription in 3 equal installments, how much is each payment?

4. Pat purchases 3.4 pounds of hamburger meat at $0.97 per pound. What is the total cost of the hamburger meat?

5. The White family took $780 cash with them on vacation. At the end of their vacation, they had $12.37 left. How much cash did they spend on vacation?

6. Acer Middle School spent $1,528.32 on 45 math books. How much did each book cost?

7. The Junior Beta Club needs to raise $1,728.32 to go to a national convention. If they decide to sell candy bars at $1.75 each, how many will they need to sell to meet their goal?

8. Fleta owns a candy store. On Monday, she sold 7.2 pounds of chocolate, 14.7 pounds of jelly beans, 2.5 pounds of sour snaps, and 9.03 pounds of yogurt-covered raisins. How many pounds of candy did she sell total?

9. Randal purchased a rare coin collection for $1,505.18. He sold it at auction for $2,600. How much money did he make on the coins?

10. A leather jacket that normally sells for $299.99 is on sale now for $186.14. How much can you save if you buy it now?

11. At the movies, Gigi buys 0.8 pounds of candy priced at $2.15 per pound. How much did she spend on candy?

12. George has $5.00 to spend on candy. If each candy bar costs $0.72, how many bars can he buy?

Copyright © American Book Company

Chapter 3 Word Problems

3.4 Estimating Rational Numbers

Some problems require an estimated solution. In order to have enough supplies to complete a project, you may need to buy more materials than you actually need. For example, a piece of plywood measures four feet wide by eight feet long. The project requires 61 feet^2 of plywood. One piece of plywood is 32 feet^2. You will need to estimate how many sheets of plywood to buy for your project.

The best approach to finding estimates is to round off all numbers in the problem, then solve. Round down for numbers ending 0–4, and round up for numbers ending in 5–9. In the case of the project above, round off the sheet of plywood to 30 feet^2. Take the 60 feet^2 needed for the project divided by the 30 feet^2 per sheet of plywood. The estimated number of sheets of plywood needed is two sheets.

In the following problems, be sure to round your answer up to the next whole number to find the correct solution.

1. The Riveras have four dogs that together eat 11 pounds of dog food per week. The dog food comes in 20 pound bags. Estimate the number of bags the Rivera's dogs will eat in 10 weeks.

2. Eliza needs $\frac{7}{8}$ yards of fabric for each skirt she is making for her 7 cousins. Estimate how much fabric Eliza will need to buy.

3. Estimate 45.6×2.1 to the nearest whole number.

4. Estimate $11\frac{4}{5} + 9\frac{1}{10}$ to the nearest whole number.

5. Bernie bags groceries using bags that come 300 per box. On one shift, Bernie and the other baggers used 1, 117 bags. Estimate how many boxes of bags Bernie and the other baggers used on the one shift.

6. Estimate to the nearest hundred $979 + 821 + 388$.

7. Estimate to the nearest million $1, 007, 331 + 12, 879, 211 - 5, 602, 001$.

8. Estimate to the nearest ten thousand $4, 973 \times 24$.

9. Marie is helping her mom make cookies for the school bake sale. Each batch makes 38 cookies and they agreed to make 200 cookies. How many batches will Marie and her mom have to make?

10. Estimate to the nearest whole number $735.57 \div 32.07$.

28   Copyright © American Book Company

3.5 Exact Versus Estimated Answers

Some problems require an estimated solution. In order to have enough product to complete a job, you often must buy more materials than you actually need. **In the following problems, be sure to round your answer up to the next whole number to find the correct solution.**

1. Endicott Publishing receives an order for 650 books. Each shipping box holds 30 books. How many boxes do the packers need to ship the order?

2. Rebecca's 250 chickens laid 214 eggs in the last 2 days. How many egg cartons holding one dozen eggs will be needed to hold all the eggs?

3. Mario's Italian restaurant uses $1\frac{1}{5}$ quarts of olive oil every day. The restaurant is open 7 days a week. For the month of September, how many gallons should the cooks order to have enough?

4. Cherokee High School is taking 297 students and 26 chaperones on a field trip. Each bus holds 46 persons. How many buses will the school need?

5. Lucille volunteers to hem 8 choir robes that came in too long. Each robe is 6 feet around the bottom. Hemming tape comes three yards to a pack. How many packs will Fran need to buy to go around all the robes?

6. Tonya is making matching vests for a children's choir. Each vest has 5 buttons on it, and there are 29 children in the choir. The button she picks comes 8 buttons to a card. How many cards of buttons does she need?

7. Tiffany is making the bread for a banquet. She needs to make 5 batches with $1\frac{3}{4}$ lb of flour in each batch. How many 10 lb bags of flour will she need to buy?

8. The homeless shelter is distributing 350 sandwiches per day to hungry guests. It takes one foot of plastic wrap to wrap each sandwich. There are 125 feet of plastic wrap per box. How many boxes will Linda need to buy to have enough plastic wrap for the week?

9. An advertising company has 12 different kinds of one-page flyers. The company needs 100 copies of each kind of flyer. How many reams of paper will the company need to produce the flyers? One ream equals 500 sheets of paper.

Other problems require an exact solution. For example, if you go to the grocery store, you can't just estimate how much you owe them. You have to find and pay the exact value.
State whether the problem requires an estimated or an exact solution.

10. You need to repay a friend for money you borrowed last week.

11. A recipe calls for a dash of pepper.

12. You have to pay the cashier for the food you ordered.

13. The number of uniforms for a cheerleading team.

Copyright © American Book Company

Chapter 3   Word Problems

3.6   Reasonable Solutions

A reasonable solution is found by taking what you know about estimation and applying it to real-world situations.

Example 1:   Sarah's school is having an assembly on Thursday. If there are 400 students at her school and the teachers are lining chairs up in rows of 15, is it reasonable to say it will take 50 rows to hold all the students? If so, how many extra seats will there be? If not, how many rows of 15 do they need?

Solution: 50 rows is not a reasonable estimation. 50 rows of 15 chairs will hold 750 students. Since there are only 400 students in Sarah's school, the number of rows can be reduced to 27 and if there are 27 rows of 15 seats, there will only be 5 empty seats.

Determine if the estimation is appropriate. If it is, state whether or not there will be any excess. If it is not appropriate, find a reasonable solution.

1. Russell is having a party. If he invites 55 people and plans on serving snacks, is it reasonable for him to go to Cost Club and buy a set of 50 plates, cups, forks, knives, and spoons (assume each person will only use one of each)? Or should he get the set of 100?

2. Brittany is planning her wedding. If she sent out 82 invitations and each person brings 1 guest (assuming everyone invited attends), would 200 chairs be reasonable for her guests at the reception? (The chairs only come in packages of 25.)

3. Sharon just got a new job and she starts Monday at 9 AM. Her office is 35 minutes away and traffic is usually very light. Since it's her first day, she'd like to be at least 15 minutes early. Can she make it in time if she leaves at 8:30 AM?

4. Eric is dog-sitting for his neighbor, Mr. McDonald. His dog, Nala, eats $\frac{1}{2}$ lb. of food twice a day. If Eric is watching Nala for 4 days, should Mr. McDonald send a 15 lb. bag to Eric's? Bags of dog food come in 5 lb. increments (5, 10, 15, etc.).

5. Sudy is baking cookies for the Christmas Show. She signed up to bring 500 cookies. Her oven can bake 65 cookies at one time, but 10 cookies usually end up burnt. Using the given information, how many batches of cookies will Sudy have to make?

6. If Sudy's recipe calls for 3.5 cups of sugar per batch, and there are 10 cups of sugar per bag, using the information found in question 5, how many bags of sugar will she need?

7. Branden is going to a baseball game. The first thing he does when he gets there is go to the concession stand. He decides to order 2 hot dogs ($2 each), 2 bags of chips ($1), some boiled peanuts ($2.50), a candy bar ($0.50), and 2 bottles of water ($1.25 each). What is the best way for him to pay if he has a five, a ten, and a twenty?

8. Jack is paving his driveway. It takes 1,000 lb. of concrete to do a 10 ft × 10 ft section, so he decides to measure the length of his driveway. If Jack's driveway is 46 feet long and 25 feet wide, how many pounds of concrete will he have to buy if you can only purchase it by the ton (2,000 lb.)?

30   Copyright © American Book Company

Chapter 3 Review

Carefully read each of the following problems and answer.

1. The Acme Aluminum Can Company sells its aluminum cans by the pound. A case of cans weighs 14 pounds. If the Green Bean Company orders 18 cases of aluminum cans, how many pounds will they receive?

2. Alicia and her mom were on the way to the shopping mall when they were pulled over by a police officer. The officer said Alicia's mom was going too slow at -8 miles under the minimum speed of 45 miles per hour. How fast was Alicia's mom going?

3. Candace weighs 97 pounds. Her younger brother, Andrew weighs 82 pounds. How many pounds different is their weight?

4. Estimate 10.2×591 to the nearest ten.

5. Sara and Sharelle are baking cherry pies for a community supper. If each pie takes 4 cups of cherries, and a container of cherries holds 8.5 cups, how many containers of cherries will the girls need to make 10 pies?

6. Mrs. Tate brought $4\frac{1}{2}$ pounds of candy to divide among her 24 students. If the candy was divided equally, how many pounds of candy did each student receive?

7. Elenita used $2\frac{2}{5}$ yards of material to recover one dining room chair. How much material would she need to recover all eight chairs?

8. The square tiles in Mr. Cooke's math classroom measure $3\frac{1}{4}$ feet across. The students counted that the classroom was $6\frac{3}{8}$ tiles wide. How wide is Mr. Cooke's classroom?

9. The Vargas family is hiking a $24\frac{2}{3}$ mile trail. The first day, they hiked $7\frac{1}{2}$ miles. How much further do they have to go to complete the trail?

10. Margo's Mint Shop has a machine that produces 3.25 pounds of mints per hour. How many pounds of mints are produced in each 8 hour shift?

11. Carter's Junior High track team ran the first leg of a 400 meter relay race in 9.8 seconds, the second leg in 11.12 seconds, the third leg in 11.47 seconds, and the last leg in 10.32 seconds. How long did it take for them to complete the race?

12. There are 4 high schools with 500 students each in the Walman County School District. The superintendent called each principal and asked how many of their students rode the bus to school. His responses were: 0.68, 321, $\frac{5}{8}$, and $\frac{7}{13}$. Which response is the highest number?

13. Terri is packing boxes of cookies into a bigger box. A big box will hold 8 boxes of cookies. If she has 61 boxes of cookies, how many big boxes will she need?

14. Jane is sewing buttons on sweaters. She has 4 new sweaters and she is sewing 5 buttons on each sweater. If buttons come in packs of 6, how many packs will she need?

Copyright © American Book Company

31

Chapter 3   Word Problems

Chapter 3 Test

1. Chester plays football at East Washington Junior High School. In one quarter of Friday's game, Chester ran $+15$ yards, -23 yards, $+34$ yards, $+17$ yards, and -28 yards. How many yards is Chester from his starting point?

    A 117

    B 71

    C 61

    D 57

    E 15

2. Abe and Juan found three books with the information they needed for a biography report of their favorite mathematician. Book one had 326 pages in it, of which 27 pages applied to their report. Book two had 419 pages in it, of which 46 pages could be useful. Book three had 542 pages in it and 33 pages contained needed information. If the two boys split up the reading equally between them, how many applicable pages will each boy read?

    F. 53

    G. 179.5

    H. 216

    J. 262.5

    K. 643.5

3. Jena walked $\frac{1}{4}$ of a mile to a friend's house, $2\frac{3}{4}$ miles to the store, and $\frac{1}{5}$ of a mile back home. How far did Jena walk?

    A $2\frac{1}{5}$ miles

    B $3\frac{1}{5}$ miles

    C $2\frac{3}{4}$ miles

    D $2\frac{4}{5}$ miles

    E $3\frac{1}{4}$ miles

4. Estimate to the nearest hundred: $314 + 599 + 627 + 151$.

    F. $1,400$

    G. $1,500$

    H. $1,600$

    J. $1,691$

    K. $1,700$

5. The Sweetest Pets Animal Shelter uses 450 pounds of cat food and 670 pounds of dog food each week. If the food comes in 25 pound bags, how many bags of food are needed of each kind weekly?

    A 16 bags for cats, and 26 bags for dogs.

    B 18 bags for cats, and 27 bags for dogs.

    C 19 bags for cats, and 27 bags for dogs.

    D 19 bags for cats, and 29 bags for dogs.

    E 27 bags for cats, and 19 bags for dogs.

6. Charlie got 4 out of 7 right on a quiz. Rosa got 0.89 correct on her test. Francisco got 5 out of 8 right on his test. Beth got 0.87 correct on her quiz. Who earned the highest grade?

    F. Charlie

    G. Rosa

    H. Francisco

    J. Beth

    K. Both Beth and Francisco

7. The park manager determined that the width of the bird sanctuary is $\frac{247}{3}$ feet wide. *Approximately*, what is this distance?

    A 82.3 ft

    B 83 ft

    C 94 ft

    D 164.4 ft

    E 13,508 ft

32   Copyright © American Book Company

8. Gene works for his father sanding wooden rocking chairs. He earns $15.77 per chair. What is the least number of chairs he needs to sand in order to buy a portable radio/CD player for $158.46?

F. 9
G. 10
H. 11
J. 12
K. 13

9. Cory uses $3\frac{1}{3}$ gallons of paint to mark one mile of this year's spring road race. How many gallons will he use to mark the entire $5\frac{1}{2}$ mile course?

A $2\frac{1}{6}$ gallons

B $3\frac{1}{3}$ gallons

C $8\frac{5}{6}$ gallons

D $15\frac{1}{6}$ gallons

E $18\frac{1}{3}$ gallons

10. Mable had $482.34 in her checking account. After writing one check in the amount of $10.93 and one check in the amount of $154.67, the bank cleared the two checks. What is Mable's new balance in her check book, assuming she has not made any more deposits?

F. $647.94
G. $327.67
H. $424.92
J. $316.74
K. $598.42

11. If you are paying someone back for money they loaned you, you should know the

A exact amount.
B estimated amount.
C believable amount.
D reasonable amount.
E order of operations.

12. Super-X sells tires for $27.95 each. Save-Rite sells the identical tire for $19.97. How much can you save by purchasing a tire from Save-Rite?

F. $5.95
G. $7.98
H. $8.00
J. $47.92
K. $58.84

13. Major Junior High School was having their school carnival. Mothers of the students made 53 cakes for the cake walk. Some of the cakes didn't quite make it to the cake walk whole, but were partially eaten by some hungry teachers. At the end of the day there was a total of $3\frac{5}{8}$ cakes leftover. How many cakes were eaten and used in the cake walk?

A $49\frac{3}{8}$

B $56\frac{5}{8}$

C $50\frac{3}{8}$

D $49\frac{5}{8}$

E $56\frac{3}{8}$

14. Lenny is taking his child, Rebecca, and six of her friends to see Anna Cabana in concert. Lenny's car will only hold four people including the driver so he had to ask another parent to drive. If the other parent's car will only hold four people including the driver as well, will they have enough room, or do they need to ask a third parent to drive?

F. enough room with 0 empty seats
G. enough room with 1 empty seat
H. enough room with 2 empty seats
J. enough room with 3 empty seats
K. They need to ask another parent to drive.

Copyright © American Book Company

Chapter 4
Percents

4.1 Changing Percents to Decimals and Decimals to Percents

To change a **percent** to a **decimal**, move the **decimal** point two places to the left, and drop the **percent** sign. If there is no decimal point shown, it is understood to be after the number and before the percent sign. Sometimes you will need to add a "0". (See 8% below.)

Example 1: $23\% = 0.23$ $8\% = 0.08$ $100\% = 1$ $409\% = 4.09$
            ↑
      (decimal point)

Change the following percents to decimal numbers.

1.	35%	5.	100%	9.	70%	13.	644%	17.	22%
2.	98%	6.	62%	10.	33%	14.	12%	18.	7%
3.	9%	7.	432%	11.	800%	15.	90%	19.	500%
4.	10%	8.	19%	12.	2%	16.	14%	20.	87%

To change a decimal to a percent, move the decimal two places to the right, and add a percent sign. You may need to add a "0". (See 0.2 below.)

Example 2: $0.24 = 24\%$ $0.03 = 3\%$ $0.2 = 20\%$ $0.445 = 44.5\%$ $2.37 = 237\%$

Change the following decimal numbers to percents.

21.	0.26	25.	1.11	29.	8.52	33.	5.55	37.	3.26
22.	0.84	26.	0.214	30.	2.33	34.	4.04	38.	0.75
23.	6.52	27.	1.8	31.	0.77	35.	0.002	39.	0.1
24.	0.99	28.	0.003	32.	0.08	36.	0.67	40.	9.44

34   Copyright © American Book Company

4.2 Changing Percents to Fractions and Fractions to Percents

4.2   **Changing Percents to Fractions and Fractions to Percents**

Example 3:   Change 135% to a fraction.

Step 1:   Copy the number without the percent sign. 135 is the numerator (the top number) of the fraction.

Step 2:   Anytime you change a percent to a decimal, the denominator of the fraction is 100.

$$135\% = \frac{135}{100}$$

Step 3:   Simplify the fraction. $\frac{135}{100} = 1\frac{7}{20}$

Change the following percents to fractions and reduce.

1.	42%	5.	200%	9.	25%	13.	500%	17.	157%
2.	17%	6.	84%	10.	83%	14.	9%	18.	333%
3.	180%	7.	110%	11.	350%	15.	61%	19.	29%
4.	3%	8.	37%	12.	10%	16.	88%	20.	50%

Example 4:   Change $2\frac{3}{4}$ to a percent.

Step 1:   Change the mixed number $2\frac{3}{4}$ to an improper fraction. Multiply the denominator, 4, by the whole number 2. $2 \times 4 = 8$. Add the numerator to this total. $3 + 8 = 11$. The fraction is now $\frac{11}{4}$.

Step 2:   Divide the numerator by the denominator. $11 \div 4$

$$
\begin{array}{r}
2.75 \\
4\,\overline{\smash{)}\,11.000} \\
-\ \ \ 8 \\
\hline
30 \\
-\ \ 28 \\
\hline
20 \\
-\ \ 20 \\
\hline
0
\end{array}
$$

Step 3:   Change the decimal answer, 2.75, to a percent by moving the decimal point 2 places to the right.

$$2\frac{3}{4} = 2.75 = 275\%$$

Copyright © American Book Company   35

Chapter 4   Percents

Change the following fractions and mixed numbers to percents.

1. $\frac{1}{2}$ 5. $\frac{8}{32}$ 9. $4\frac{4}{5}$ 13. $\frac{99}{100}$ 17. $7\frac{1}{4}$

2. $\frac{9}{10}$ 6. $3\frac{7}{8}$ 10. $\frac{1}{4}$ 14. $\frac{45}{100}$ 18. $\frac{3}{8}$

3. $2\frac{1}{2}$ 7. $9\frac{3}{4}$ 11. $5\frac{3}{8}$ 15. $3\frac{1}{2}$ 19. $\frac{1}{5}$

4. $3\frac{1}{4}$ 8. $\frac{1}{16}$ 12. $12\frac{1}{2}$ 16. $\frac{7}{8}$ 20. $6\frac{1}{4}$

4.3   Percent Word Problems

Example 5:   Wayne has spent 40% of his allowance on paper for his computer printer. If Wayne spent exactly $3.60 including tax on the paper, how much is his allowance, and how much does he have left to spend?

Step 1:   Turn the percentage from the problem into a decimal. $40\% = 0.4$

Step 2:   Divide the dollar amount by the percent. $3.60 \div 0.4 = \$9$

Step 3:   Wayne's allowance is $9.00 and he has $9.00 - \$3.60 = \$5.40$ left over.

Carefully read the word problems below and solve.

1. Emily is allowed to spend 60% of her allowance. The remaining 40% she deposits into her savings account. Emily deposited $2.00 into her savings account this week. What amount does Emily get each week for her allowance?

2. Dashon walks 30 miles a week for exercise. He has completed 40% of his weekly goal. How many more miles does Dashon have to walk to meet his goal?

3. Gretchen wants to make a quilt for her bed. She has completed 60% of the quilt. When completed, the quilt will be made up of 90 pieces. How many pieces does Gretchen have left to sew?

4. Hank is buying a shirt that originally cost $25.00. It was on sale two weeks ago for 10% off. Today, it is an additional 20% off the original price. How much is the shirt now?

5. Alicia is walking around the shopping mall with her mom to find the best price on a new blouse. One store has a blouse she wants for 25% off the original price of $40.00. Another store has an almost identical blouse for 30% off the original price of $50.00. Which blouse will cost less, the one priced 25% off or the one priced 30% off?

36   Copyright © American Book Company

4.4   Finding the Percent of the Total

Example 6:   800 people came to the high school football game. Sixty-five percent of the people came to cheer for the home team. How many people came to cheer for the home team?

Step 1:   Change 65% to a decimal.

Step 2:   Multiply the percent by the number of people who attended the game.
$800 \times 0.65 = 520$

520 people came to cheer the home team.

Carefully read the problems below and solve.

1. Lyla baked 200 cupcakes for the school bake sale. She sold 80% of them. How many did she sell?

2. Hill's Department Store had a one day sale on footballs just before the start of the fall football season. They started the day with 1,000 footballs, and sold 87 percent of the footballs. How many footballs did the store have left at the end of the day?

3. The kids at Milridge Middle School were trying to earn a special field trip day by reading 10,000 books. By January, they school had read 6,500 books. What percent of their goal had the school reached so far?

4. Cary's dad gave her $20.00 to spend on shoes at the mall. Cary spent 85% of the money on one pair of shoes. How much money does Cary's dad get back from his $20.00?

5. The lunchroom at Prairie Middle School sold 90% of the burger meat they ordered for the week. If Mrs. Halgurst ordered 900 pounds of meat for a week's worth of burgers, how many pounds of meat did the lunchroom sell?

6. Thirty percent of sixty students polled said dogs were their favorite pets. How many of the sixty students preferred dogs?

7. Forty-five percent of the same sixty students polled said cats were their favorite pets. How many of the sixty students preferred cats?

8. Hilda works at a candy store and sold 80 pounds of peanut butter candies in one day. Computer totals showed that of the 120 customers that day, 40% bought peanut butter candies. How many customers purchased peanut butter candies that day?

9. Eighty-two percent of 300 boys polled said that they liked to play outdoors. How many boys liked to play outdoors?

Copyright © American Book Company

Chapter 4   Percents

4.5   Percent Increase or Decrease

When finding the percent increase or decrease in a word problem, you must first identify the **original** number and then the **change**. The **original** number is identified by the value that takes place first the word problem. It may or may not be mentioned first in the problem. You will need to study the word problem carefully to identify the values.

The **change** is the difference between the earlier value and the current value in the word problem. This can be figured out by simply subtracting the two numbers.

To find the percent increase or decrease simply divide the **change** by the **original** value.

$$\text{Percent Change} = \frac{\text{Change}}{\text{Original}}$$

Example 7:   The population of Wilkinsville was $3,200$ one year and increased to $3,520$ the next year. What is the percent **increase** of the population of Wilkinsville?

Step 1:   Find the difference in the two numbers. This equals the change in population.
$3,520 - 3,200 = 320$

Step 2:   Divide the difference by the original population. $320 \div 3,200 = 0.10$

Step 3:   Change the answer to a percent. $0.10 = 10\%$
The population increased by 10%.

Example 8:   Mr. Hunter had 25 students in his 4th period math class. The next day, he had only 20 students. What is the percent **decrease** in the number of students in Mr. Hunter's science class?

Step 1:   Find the difference in the two numbers. This equals the change in the number of students.
$25 - 20 = 5$

Step 2:   Divide the difference by the original number of students.
$5 \div 25 = 0.20$

Step 3:   Change the answer to a percent. $0.20 = 20\%$
The number of students in Mr. Hunter's class decreased by 20%.

38   Copyright © American Book Company

4.5   Percent Increase or Decrease

Carefully read the problems below and solve.

1. The middle school's baseball team, the Badgers, won 20 games last year. This year they won 22 games. What is the percent increase of games won from this year to last year?

2. Tony bought a printer on sale for $45.00. The original price was $60.00. What is the percent decrease in price on the printer?

3. If a loaf of bread was priced $1.50 on one day and $1.20 the next, what is the percent decrease of the price of the loaf of bread?

4. The Toys of Life store had 240 customers on October 24th. They had 264 customers on October 25th. What was the percent increase in customers from October 24th to October 25th?

5. Dr. Baler's veterinary practice is booming. Last year he had 1,500 appointments. This year he had 2,250 appointments and had to hire more help. What is the percent increase in appointments from this year to last year?

6. Lisa's little brother weighed 8 pounds when he was born. He now weighs 36 pounds. What is the percent increase in weight of the little boy?

7. John's dog, Drooley, is old and ill. He used to weigh 70 pounds and now weighs 56 pounds. What is the percent decrease in weight of Drooley?

8. Rita Rose had 120 buttons in her button collection. She now has 150 buttons. What is the percent increase in the number of buttons in Rita Rose's collection?

9. Matthew had $50.00 in his special hiding place. He decides to spend $15.00 on his mom's birthday present. What percent decrease in money will this be for Matthew?

10. Farmer Jake harvested 2,500 pounds of carrots this fall. He sold them out of the back of his truck and brought home only 250 pounds after a full day of selling. What is the percent decrease in the number of carrots Farmer Jake has left?

Copyright © American Book Company

Chapter 4   Percents

4.6   Finding the Amount of Discount

Prices are sometimes marked a percent off to sell remaining inventory in a hurry. The amount you save is called the **discount**.

Example 9:   A chair that sells for $159.00 is on sale for 30% off. How much can you save if you buy the chair on sale?

   Step 1:   Change the percent of the discount to a decimal. 30% = 0.30

   Step 2:   Multiply the original price of the chair by the discount.
   $159.00 × 0.30 = $47.70. You will save $47.70 if you buy the chair on sale.

Figure the amount of discount on each of the examples below.

1. Hamburger is on sale for 20% off the normal $3.70 a pound price. How much can you save per pound buying the meat on sale?

2. Hardcover books are on sale for 15% off. Louise has chosen a hardcover book that normally sells for $20.00. How much will Louise save?

3. Alex chose a pair of jeans that sells for $19.50, but today all jeans are on sale for 10% off. How much will Alex save?

4. Dolly's Dress Shop is having a sale on dresses - buy one, get the second for half off (50%). Madison finds two dresses that normally sell for $50.00 each. How much will she save if she buys the two dresses during this special sale?

5. Henry finds a sale on tires, buy 4 - only pay for three! Henry buys four tires that normally sell for $40.00 each. How much will Henry pay for the 4 tires?

6. At a back to school sale for computers, Irwin buys one that normally costs $599.00. He receives the sale price at 10% off. How much does Irwin save on the computer. (Assume the sales tax is included in the original price of $599.00).

7. Mrs. Barkins is buying shoes for all six of her children. The total of the six pairs comes to $138.00. This week, the store is running a sale - 30% off all children's shoes. How much does she save buying the six pairs of shoes this week?

8. Mary Ellen has a coupon to get 25% off any sweater at the Yellow Bee Dress Shop. Mary Ellen chooses a sweater that costs $44.80, including tax. How much will Mary Ellen save with her coupon?

40   Copyright © American Book Company

4.7 Finding the Discounted Sale Price

To find the discounted sale price, you must go one step further than shown on the previous page. Read the example below to learn how to figure **discount** prices.

Example 10: A $62.00 chair is on sale for 20% off. How much will it cost if I buy it now?

Step 1: Change 20% to a decimal.

$$20\% = 0.2$$

Step 2: Multiply the original price by the discount.

$$
\begin{array}{r}
\textbf{ORIGINAL PRICE} \\
\times \quad \textbf{\% DISCOUNT} \\
\hline
\textbf{SAVINGS}
\end{array}
\qquad
\begin{array}{r}
\$62.00 \\
\times \quad 0.2 \\
\hline
\$12.40
\end{array}
$$

Step 3: Subtract the savings amount from the original price to find the sale price.

$$
\begin{array}{r}
\textbf{ORIGINAL PRICE} \\
- \quad \textbf{SAVINGS} \\
\hline
\textbf{SALE PRICE}
\end{array}
\qquad
\begin{array}{r}
\$62.00 \\
- \quad 12.40 \\
\hline
\$49.60
\end{array}
$$

Figure the sale price of the items below. The first one is done for you.

ITEM	PRICE	%OFF	MULTIPLY	SUBTRACT	SALE PRICE
1. pen	$2.50	30%	$2.50 \times 0.3 = \$0.75$	$2.50 - 0.75 = 1.75$	$1.75
2. recliner	$420	50%			
3. juicer	$75	25%			
4. blanket	$16	30%			
5. earrings	$1.80	15%			
6. figurine	$12	20%			
7. boots	$84	10%			
8. calculator	$12	10%			
9. candle	$7.95	70%			
10. camera	$218	5%			
11. DVD player	$99.95	10%			
12. video game	$45	25%			

Copyright © American Book Company

41

Chapter 4   Percents

Chapter 4 Review

1. Change 41% to a decimal.

2. Change 0.315 to a percent.

3. Change 923% to a decimal.

4. Change 4.67 to a percent.

5. Change 92% to a fraction.

6. Change $\frac{2}{5}$ to a percent.

7. Change 478% to a fraction.

8. Change $2\frac{1}{2}$ to a percent.

9. Ayana practices her violin 8 hours every week. So far this week, she has practiced 6 hours. What percent of Ayana's practice has she completed for the week?

10. Jeremy bought a pair of pants on a 40% off sale. The original price of the pants was $18.00 including tax. How much did Jeremy pay for the pants?

11. One hundred fifty students at Park City Middle School were polled and asked what is their favorite kind of movie. Twenty-two percent of the students polled said their favorite kind of movie was Science Fiction. How many students preferred Science Fiction movies?

12. The Wolverines basketball team won 12 games last season. This year, the team won 15 games. What percent increase did the Wolverines win this year over last year?

13. Tim and his dad wanted to buy new T-shirts. There was a sale at the T-Shirt Factory Store. If you buy at least 3 T-shirts, you got 35% off your entire T-shirt purchase. Tim and his dad together purchased a total of 4 T-shirts for $8.00 each. How much did they save?

14. Laura has a coupon to get $1.00 off a $4.00 bottle of laundry detergent. What percent will Laura save?

Chapter 4 Test

1. Change 94% to a decimal.

   A. 0.94
   B. 9.4
   C. 94
   D. 0.094
   E. 0.904

2. Change 6.22 to a percent.

   F. 6.22%
   G. 62.2%
   H. 622%
   J. 0.622%
   K. 0.0622

3. Change 40% to a fraction.

   A. $\frac{1}{20}$
   B. $\frac{1}{40}$
   C. $4\frac{1}{10}$
   D. $\frac{2}{5}$
   E. $\frac{40}{10}$

4. Melea bought a blouse that was originally marked $28.00 for 20% off. How much did Melea pay for the blouse?

   F. $20.00
   G. $5.60
   H. $22.40
   J. $20.40
   K. 0.0622

42   Copyright © American Book Company

Chapter 4 Test

5. Liz is collecting all 24 dolls in a special collection. So far, she has received 18 of the dolls for her birthdays and special occasions. What percent of the collection does Liz have so far?

   A. 75%

   B. 18%

   C. 24%

   D. 50%

   E. 0.904

6. Colin's puppy, Squeeker, weighed 1.5 pounds at his first vet visit. When Squeeker was a year old, he weighed 7.5 pounds. What percent did Squeeker's weight increase between the two visits?

   F. 500%

   G. 300%

   H. 400%

   J. 600%

   K. 0.0622

7. Breona brought 4 pounds of cookies to a slumber party. When the girls were done eating, there was 1 pound of cookies left. What percent did the girls at the slumber party eat?

   A. 75%
   B. 50%
   C. 25%
   D. 7.5%
   E. 0.904

8. What is 86% of 200?

   F. 172
   G. 86
   H. 286
   J. 114
   K. 0.0622

9. Mr. Allen had 8 hamsters in cages in his science class. The next year, Mr. Allen had 10 hamsters. What percent increase in hamsters did Mr. Allen have?

   A. 18%
   B. 20%
   C. 25%
   D. 2%
   E. 0.904

10. A hat costs $17.00 and is on sale for 30% off. How much will be saved?

   F. $3.40

   G. $51.00

   H. $34.00

   J. $5.10

   K. 0.0622

11. Change $\frac{5}{8}$ to a percent.

   A. 6.25%
   B. 0.625%
   C. 62.5%
   D. 625%
   E. 0.904

12. Change 0.01 to a fraction.

   F. $\frac{1}{10}$

   G. $\frac{10}{100}$

   H. $\frac{10}{1}$

   J. $\frac{1}{100}$

   K. $\frac{1}{1,000}$

Copyright © American Book Company

Chapter 5
Exponents and Roots

5.1   Understanding Exponents

Exponents are a shorter way to express multiplying a number by itself one or more times. For example, a math problem may need to multiply 4×4 or $7 \times 7 \times 7 \times 7$. A shorter way to write these math problems is 4^2 and 7^4. The first number is the **base**. The small, raised number is called the **exponent** or **power**. The exponent tells how many times the base is multiplied by itself.

Exponents are read as: 4^2 is four to the second power, or four squared

6^3 is six to the third power, or six cubed

9^7 is nine to the seventh power

118^{27} is one hundred eighteen to the twenty-seventh power

Example 1: 6^3 ◄— **exponent (or power)**   This means multiply by 6 three times: $6 \times 6 \times 6$
   ◄— **base**

Example 2:   **Negative numbers can be raised to exponents also.**
An **even** exponent will give a **positive** answer:
$(-2)^2 = (-2) \times (-2) = 4$
An **odd** exponent will give a **negative** answer:
$(-2)^3 = (-2) \times (-2) \times (-2) = (-8)$

You also need to know two special properties of exponents:

1.   Any base number raised to the exponent of 1 equals the base number.
2.   Any base number raised to the exponent of 0 equals 1.

Example 3: $4^1 = 4$ $10^1 = 10$ $25^1 = 25$ $4^0 = 1$ $10^0 = 1$ $25^0 = 1$

44   Copyright © American Book Company

5.1   Understanding Exponents

Rewrite the following problems using exponents. $(3 \times 3 \times 3 \times 3 = 3^4)$

1. $12 \times 12 \times 12 \times 12 \times 12$

2. $111 \times 111 \times 111$

3. $42 \times 42 \times 42 \times 42$

4. 2×2

5. $7 \times 7 \times 7 \times 7 \times 7 \times 7 \times 7$

6. $33 \times 33 \times 33$

7. $14 \times 14 \times 14 \times 14$

8. $1 \times 1 \times 1 \times 1 \times 1 \times 1$

9. $52 \times 52 \times 52$

10. $5,687 \times 5,687 \times 5,687$

11. 45×45

12. $8 \times 8 \times 8 \times 8 \times 8 \times 8 \times 8 \times 8 =$

Find what product each number with an exponent represents.
$(2^3 = 2 \times 2 \times 2 = 8)$

13. $(-2)^3$

14. 4^2

15. 100^1

16. 9^4

17. $4,724^0$

18. 13^2

19. $(-24)^2$

20. 4^5

21. 3^3

22. 1^0

23. 6^2

24. $71,852^1$

25. $(-5)^3$

26. 5^3

27. 8^3

Express each of the following numbers as a base with an exponent. $(4 = 2 \times 2 = 2^2)$

28. 25

29. 64

30. 144

31. 4

32. $10,000$

33. 8

34. 81

35. 27

36. 128

37. 100

38. 625

39. 36

Copyright © American Book Company

Chapter 5   Exponents and Roots

5.2   Multiplying Exponents with the Same Base

To multiply two expressions with the same base, add the exponents together and keep the base the same.

Example 4: $3^4 \times 3^2 = 3^{4+2} = 3^6$

Example 5: $2a^2 \times 4a^2 = 8a^{2+2} = 8a^4$
Notice that only the a's are raised to a power and not the 2 or the 4.

Simplify each of the expressions below.

1. $x^2 \times x^7$

2. $8^3 \times 8^4$

3. $4y^2 \times y^3$

4. $3^8 \times 3^1$

5. $2a^6 \times 3a^2$

6. $3x^2 \times x^1$

7. $6^2 \times 6^3$

8. $7^2 \times 7^2$

9. $4^2 \times 4^5$

10. $2x \times x^3$

11. $a^3 \times 4a^2$

12. $8^2 \times 8^6$

5.3   Multiplying Exponents Raised to an Exponent

If a power is raised to another power, multiply the exponents together and keep the base the same.

Example 6: $(4^3)^4 = 4^{3\times4} = 4^{12}$

Example 7: $(x^3)^2 = x^{3\times2} = x^6$

Simplify each of the expressions below.

1. $(8^2)^3$

2. $(3^4)^3$

3. $(y^3)^5$

4. $(2^4)^1$

5. $(x^2)^2$

6. $(10^3)^2$

7. $(a^1)^3$

8. $(z^5)^2$

9. $(4^3)^3$

10. $(9^4)^5$

11. $(12^2)^6$

12. $(1^3)^1$

46   Copyright © American Book Company

5.4 More Multiplying Exponents

If a product in parentheses is raised to a power, then each factor is raised to the power when parentheses are eliminated.

Example 8: $(3 \times 5)^2 = 3^2 \times 5^2 = 9 \times 25 = 225$

Example 9: $(4a)^2 = 4^2 \times a^2 = 16a^2$

Example 10: $(9x^7)^2 = 9^2 x^{14} = 81x^{14}$

Simplify each of the expressions below.

1. $(4^2)^2$

2. $(2b^3)^2$

3. $(3a^3)^2$

4. $(2^5)^2$

5. $(2 \times 3)^2$

6. $(5x^2)^3$

7. $(7w^5)^2$

8. $(8p^2)^0$

9. $(4x^2)^3$

10. $(6 \times 2)^2$

11. $(3y)^3$

12. $(2^2)^3$

13. $(2 \times 3)^3$

14. $(4^3)^0$

15. $(3z^8)^3$

16. $(5n^3)^3$

17. $(4 \times 4)^2$

18. $(4x^5)^3$

19. $(5a^3)^2$

20. $(8b^8)^2$

21. 2×2^3

22. $(7x^9)^2$

23. $(2 \times 4)^2$

24. $(15z^5)^2$

25. $(9b^5)^2$

26. $(3a^3)^3$

27. $(6t^4)^2$

28. $(3y^3)^2$

29. (8×5^2)

30. $(2^2)^2$

31. $(8b^5)^3$

32. $(12c^3)^2$

Answer each question below.

33. Is x^6 equal to $(x^2)^3$?

34. Is x^7 equal to $(7x)^7$?

35. Is $4x^8$ equal to $(2x^4)^2$?

36. Is $13x^9$ equal to $13(x^3)^3$?

37. If x^4 is equal to 16, what is $(x^2)^2$ equal to?

38. If $2x^3$ is equal to 54, what is $2 \cdot x^3$ equal to?

39. If $7x^2$ is equal to 63, what is $7 \cdot x^2$ equal to?

40. If $10x^{12}$ is equal to 10, what is $10x^{11}$ equal to?

Chapter 5   Exponents and Roots

5.5   Negative Exponents

Expressions can also have negative exponents. Negative exponents do not indicate negative numbers. They indicate **reciprocals**. The **reciprocal** of a number is one divided by that number. For example, the reciprocal of 2 is $\frac{1}{2}$. A number multiplied by its reciprocal is equal to 1. If the negative exponent is in the bottom of a fraction, the reciprocal will put the expression on the top of the fraction without a negative sign.

Example 11: $4^{-2} = \dfrac{1}{4^2} = \dfrac{1}{16}$

Example 12: $5b^{-3} = 5 \times \dfrac{1}{b^3} = \dfrac{5}{b^3}$ Notice that the 5 is not raised to the -3 power, only the b.

Example 13: $\dfrac{4}{3y^{-3}} = \dfrac{4y^3}{3}$ Notice that the 3 is not raised to the -3 power, only the y.

Rewrite using only positive exponents.

1. p^{-3}

2. $\dfrac{x^{-2}}{4}$

3. $8y^{-9}$

4. $\dfrac{3m^{-2}}{2}$

5. $7z^{-4}$

6. $10m^{-3}$

7. $\dfrac{p^{-1}}{12}$

8. $17m^{-8}$

9. $6x^{-6}$

10. $\dfrac{3}{2t^{-8}}$

11. $\dfrac{n^{-7}}{4}$

12. $12z^{-9}$

13. r^{-2}

14. $-3x^{-5}$

15. $\dfrac{8m^{-2}}{7}$

5.6   Multiplying with Negative Exponents

Multiplying with negative exponents follows the same rules as multiplying with positive exponents.

Example 14: $5^3 \cdot 5^{-4} = 5^{3+(-4)} = 5^{-1} = \dfrac{1}{5}$

Example 15: $(2a \times 4)^{-2} = (8a)^{-2} = \dfrac{1}{(8a)^2} = \dfrac{1}{64a^2}$

Example 16: $(8z^3)^{-2} = 8^{-2}z^{-6} = \dfrac{1}{8^2 z^6} = \dfrac{1}{64z^6}$

Simplify the following. Answers should <u>not</u> have any negative exponents.

1. $3^{-3} \cdot 3^6$

2. $9^{-4} \times 9^2$

3. $6^{-5} \cdot 6^7$

4. $2^4 \cdot 2^{-5}$

5. $2^8 \times 2^{-11}$

6. $12^4 \times 12^{-6}$

7. $4^{-7} \cdot 4^{10}$

8. $(10^2 \cdot 10^{-3})^2$

9. $4^{-2} \cdot 4^6$

10. $(7x^5)^{-4}$

11. $(5z^3)^{-6}$

12. $7^{-2} \times 7^{-2}$

13. $(6p^4)^{-3}$

14. $10^{-10} \cdot 10^8$

15. $(3^3 \times 3^{-2})^{-2}$

16. $(2y)^{-4}$

Copyright © American Book Company

5.7   Dividing with Exponents

5.7   Dividing with Exponents

Exponents that have the same base can also be divided.

Example 17: $\dfrac{7^6}{7^4}$ This problem means $7^6 \div 7^4$. Let us look at 2 ways to solve this problem.

Solution 1: $\dfrac{7^6}{7^4} = \dfrac{7 \cdot 7 \cdot 7 \cdot 7 \cdot 7 \cdot 7}{7 \cdot 7 \cdot 7 \cdot 7} = 7 \cdot 7 = 49$ First, rewrite the fraction with the exponents in expanded form, simplify, and then multiply.

Solution 2: $\dfrac{7^6}{7^4} = 7^{6-4} = 7^2 = 49$ A quick way to simplify this same problem is to subtract the exponents. **When dividing exponents with the same base, subtract the exponents.**

Example 18: $\dfrac{(3x)^{-4}}{2x^3}$

Step 1: $(3x)^{-4} = \dfrac{1}{(3x)^4} = \dfrac{1}{3^4 x^4}$ Remove the parentheses from the top of the fraction.

Step 2: $\dfrac{1}{3^4 x^4 \cdot 2x^3} = \dfrac{1}{162x^7}$ The bottom of the fraction remains the same, so multiply the two together and simplify.

Example 19: $\dfrac{2^4}{2^7}$

Solution: $\dfrac{2^4}{2^7} = \dfrac{2 \cdot 2 \cdot 2 \cdot 2}{2 \cdot 2 \cdot 2 \cdot 2 \cdot 2 \cdot 2 \cdot 2} = \dfrac{1}{2 \cdot 2 \cdot 2} = \dfrac{1}{8}$ *PEMDAS*

Simplify the problems below. You may be able to cancel. Be sure to follow order of operations. Remove parentheses before canceling.

1. $\dfrac{4^7}{4^5}$

2. $\dfrac{y^4}{y^3}$

3. $\dfrac{5^6}{5^3}$

4. $\dfrac{(2^2)^3}{2^5}$

5. $\dfrac{(7^2)^4}{7^5}$

6. $\dfrac{14^3}{14^2}$

7. $\dfrac{8^3}{8^5}$

8. $\dfrac{5^{12}}{5^{10}}$

9. $\dfrac{10^8}{10^5}$

10. $\dfrac{4^3}{4^5}$

11. $\dfrac{y^2}{(y^3)^4}$

12. $\dfrac{17^{-5}}{17^{-4}}$

13. $\dfrac{3^{11}}{3^9}$

14. $\dfrac{z^4}{(z^2)^4}$

15. $\dfrac{9^2}{9^4}$

16. $\dfrac{12^{-2}}{(12^2)^{-2}}$

17. $\dfrac{13^{10}}{13^9}$

18. $\dfrac{5^{-3}}{5^{-5}}$

19. $\dfrac{(t^2)^{-7}}{t^8}$

20. $\dfrac{7^8}{7^{10}}$

21. $\dfrac{y^3}{(y^8)^{-2}}$

22. $\dfrac{z^3}{(z^2)^3}$

23. $\dfrac{(3^2)^3}{3^5}$

24. $\dfrac{10^2}{10^{-3}}$

Copyright © American Book Company

Chapter 5   Exponents and Roots

5.8   Order of Operations

In long math problems with $+$, $-$, $\times$, $\div$, $()$, and exponents in them, you have to know what to do first. Without following the same rules, you could get different answers. If you memorize the silly sentence, Please Excuse My Dear Aunt Sally, you can memorize the order you must follow.

Please   "P" stands for parentheses. You must eliminate parentheses first.
Examples: $3(1+4) = 3(5) = 15$
$6(10-6) = 6(4) = 24$

Excuse   "E" stands for exponents. You must eliminate exponents next.
Example: $4^2 = 4 \times 4 = 16$

My Dear   "M" stands for multiply. "D" stands for divide. Start on the left of the equation and perform all multiplications and divisions in the order in which they appear.

Aunt Sally   "A" stands for add. "S" stands for subtract. Start on the left and perform all additions and subtractions in the order they appear.

Example 20: $12 \div 2(6-3) + 3^2 - 1$

Please	Eliminate **parentheses**. $6 - 3 = 3$ so now we have	$12 \div 2 \times 3 + 3^2 - 1$
Excuse	Eliminate **exponents**. $3^2 = 9$ so now we have	$12 \div 2 \times 3 + 9 - 1$
My Dear	**Multiply** and **divide** next in order from left to right.	$12 \div 2 = 6$ then $6 \times 3 = 18$
Aunt Sally	Last, we **add** and **subtract** in order from left to right.	$18 + 9 - 1 = 26$

Simplify the following problems.

1. $6 + 9 \times 2 - 4$

2. $3(4+2) - 6^2$

3. $3(6-3) - 2^3$

4. $49 \div 7 - 3 \times 3$

5. $10 \times 4 - (7-2)$

6. $2 \times 3 \div 6 \times 4$

7. $4^3 \div 8(4+2)$

8. $7 + 8(14-6) \div 4$

9. $(2+8-12) \times 4$

10. $4(8-13) \times 4$

11. $8 + 4^2 \times 2 - 6$

12. $3^2(4+6) + 3$

13. $(12-6) + 27 \div 3^2$

14. $82^0 - 1 + 4 \div 2^2$

15. $1 - (2-3) + 8$

16. $-4\{18 - (4 + 2 \times 6)\}$

17. $18 \div (6+3) - 12$

18. $10^2 + 3^3 - 2 \times 3$

19. $4^2 + (7+2) \div 3$

20. $7 \times 4 - 9 \div 3$

50   Copyright © American Book Company

5.8   Order of Operations

When a problem has a fraction bar, simplify the top of the fraction (numerator) and the bottom of the fraction (denominator) separately using the rules for order of operations. Treat the top and bottom as if they were separate problems. Then reduce the fraction to lowest terms.

Example 21: $\dfrac{2(4-3)-6}{5^2+3(2+1)}$

Please	Eliminate **parentheses**. $(4-3)=1$ and $(2+1)=3$	$\dfrac{2 \times 1 - 6}{5^2 + 3 \times 3}$
Excuse	Eliminate **exponents**. $5^2 = 25$	$\dfrac{2 \times 1 - 6}{25 + 3 \times 3}$
My **D**ear	**Multiply** and **divide** in the numerator and denominator separately. $3 \times 3 = 9$ and $2 \times 1 = 2$	$\dfrac{2 - 6}{25 + 9}$
Aunt **S**ally	**Add** and **subtract** in the numerator and denominator separately. $2 - 6 = -4$ and $25 + 9 = 34$	$\dfrac{-4}{34}$

Now simplify the fraction to lowest terms. $\dfrac{-4}{34} = \dfrac{-2}{17}$

Simplify the following problems.

PEMDAS

1. $\dfrac{2^2+4}{5+3(8+1)}$

2. $\dfrac{8^2-(4+11)}{4^2-3^2}$

3. $\dfrac{5-2(4-3)}{2(1-8)}$

4. $\dfrac{10+(2-4)}{4(2+6)-2^2}$

5. $\dfrac{3^3-8(1+2)}{-10-(3+8)}$

6. $\dfrac{(9-3)+3^2}{-5-2(4+1)}$

7. $\dfrac{16-3(10-6)}{(13+15)-5^2}$

8. $\dfrac{(2-5)-11}{12-2(3+1)}$

9. $\dfrac{7+(8-16)}{6^2-5^2}$

10. $\dfrac{16-(12-3)}{8(2+3)-5}$

11. $\dfrac{-3(9-7)}{7+9-2^3}$

12. $\dfrac{4-(2+7)}{13+(6-9)}$

13. $\dfrac{5(3-8)-2^2}{7-3(6+1)}$

14. $\dfrac{3(3-8)+5}{8^2-(5+9)}$

15. $\dfrac{6^2-4(7+3)}{8+(9-3)}$

Copyright © American Book Company

51

Chapter 5   Exponents and Roots

5.9   Scientific Notation

Mathematicians use **scientific notation** to express very large and very small numbers. Scientific notation expresses a number in the following form:

$$x.xx \times 10^x$$

only one digit right before the decimal

multiplied by a multiple of ten

remaining digits not ending in zeros after the decimal

5.10   Using Scientific Notation for Large Numbers

Scientific notation simplifies very large numbers that have many zeros. For example, Pluto averages a distance of $5,900,000,000$ kilometers from the sun. In scientific notation, a decimal is inserted after the first digit (5.). The rest of the digits are copied except for the zeros at the end, (5.9), and the result is multiplied by 10^9. The exponent equals the total number of digits in the original number minus 1 or the number of spaces the decimal point moved.

$5,900,000,000 = 5.9 \times 10^9$ The following are more examples:

Example 22: $32,560,000,000 = 3.256 \times 10^{10}$ $5,060,000 = 5.06 \times 10^6$

decimal moves 10 spaces to the left

decimal moves 6 spaces to the left

Convert the following numbers to scientific notation.

1. $4,230,000,000$
2. $64,300,000$
3. $951,000,000,000$
4. $12,300$

5. $20,350,000,000$
6. $9,000$
7. $450,000,000,000$
8. $6,200$

9. $87,000,000$
10. $105,000,000$
11. $1,083,000,000,000$
12. $304,000$

To convert a number written in scientific notation back to conventional form, reverse the steps.

Example 23: $4.02 \times 10^5 = 4.02000 = 402,000$ Move the decimal 5 spaces to the right and add zeros.

Convert the following numbers from scientific notation to conventional numbers.

13. 6.85×10^8
14. 1.3×10^7
15. 4.908×10^4

16. 7.102×10^6
17. 2.5×10^3
18. 9.114×10^5

19. 5.87×10^7
20. 8.047×10^8
21. 3.81×10^5

22. 9.5×10^{12}
23. 1.504×10^6
24. 7.3×10^9

52   Copyright © American Book Company

5.11   Using Scientific Notation for Small Numbers

5.11   Using Scientific Notation for Small Numbers

Scientific notation also simplifies very small numbers that have many zeros. For example, the diameter of a helium atom is 0.000000000244 meters. It can be written in scientific notation as 2.44×10^{-10}. The first number is between 1 and 10, and the first number is always followed by a decimal point. The negative exponent indicates how many digits the decimal point moved to the right. The exponent is negative when the original number is less than 1. To convert small numbers to scientific notation, use the following examples.

Examples: $0.00058 = 5.8 \times 10^{-4}$ $0.00003059 = 3.059 \times 10^{-5}$

Decimal point moves 4 spaces to the right

negative exponent indicates the original number is less than 1.

decimal moves 5 spaces to the right

Convert the following numbers to scientific notation.

1. 0.00000254

2. 0.00000000508

3. 0.000008004

4. 0.00047

5. 0.000000005478

6. 0.00000059

7. 0.00000004712

8. 0.00025

9. 0.0000000501

10. 0.0000006

11. 0.0000000000875

12. 0.00004

Now convert small numbers written in scientific notation back to conventional form.

Example 24: $3.08 \times 10^{-5} = 00003.08 = 0.0000308$ Move the decimal 5 spaces to the left, and add zeros.

Convert the following numbers from scientific notation to conventional numbers.

13. 1.18×10^{-7}

14. 2.3×10^{-5}

15. 6.205×10^{-9}

16. 4.1×10^{-6}

17. 7.632×10^{-4}

18. 5.48×10^{-10}

19. 2.75×10^{-8}

20. 4.07×10^{-7}

21. 5.2×10^{-3}

22. 7.01×10^{-6}

23. 4.4×10^{-5}

24. 3.43×10^{-2}

Copyright © American Book Company

53

Chapter 5   Exponents and Roots

5.12   Square Root

Just as working with exponents is related to multiplication, finding square roots is related to division. In fact, the symbol for finding the square root of a number looks similar to a division sign. The symbol is called a **radical**. The best way to learn about square roots is to look at examples.

Example 25:   This is a square root problem: $\sqrt{64}$
It is asking, "What is the square root of 64?"
It means, "What number multiplied by itself equals 64?"
We know that $8 \times 8 = 64$, so the $\sqrt{64} = 8$.

Find the square root of the following numbers.

1. $\sqrt{49}$
2. $\sqrt{81}$
3. $\sqrt{25}$
4. $\sqrt{16}$
5. $\sqrt{121}$

6. $\sqrt{625}$
7. $\sqrt{100}$
8. $\sqrt{289}$
9. $\sqrt{196}$
10. $\sqrt{36}$

11. $\sqrt{4}$
12. $\sqrt{900}$
13. $\sqrt{64}$
14. $\sqrt{9}$
15. $\sqrt{144}$

5.13   Estimating Square Roots

Example 26:   Estimate the value of $\sqrt{3}$.

Step 1:   Estimate the value of $\sqrt{3}$ by using the square root of values that you know. $\sqrt{1}$ is 1 and $\sqrt{4}$ is 2, so the value of $\sqrt{3}$ is going to be between 1 and 2.

Step 2:   To estimate a little closer, try squaring 1.5. $1.5 \times 1.5 = 2.25$, so $\sqrt{3}$ has to be greater than 1.5. If you do further trial and error calculations, you will find that $\sqrt{3}$ is greater than 1.7 ($1.7 \times 1.7 = 2.89$) but less than 1.8 ($1.8 \times 1.8 = 3.24$).

Therefore $\sqrt{3}$ is around 1.75. It is closer to 2 than it is to 1.

Example 27:   Is $\sqrt{52}$ closer to 7 or 8? Look at the perfect square above and below 52.

To answer this question, first look at 7^2 which is equal to 49 and 8^2 which is equal to 64. Then ask yourself whether 52 is closer to 49 or 64? The answer is 49, of course. Therefore, the $\sqrt{52}$ is closer to 7 than 8.

Follow the steps above to answer the following questions. Do not use a calculator.

1. Is $\sqrt{66}$ closer to 8 or 9?
2. Is $\sqrt{27}$ closer to 5 or 6?
3. Is $\sqrt{13}$ closer to 3 or 4?
4. Is $\sqrt{78}$ closer to 8 or 9?
5. Is $\sqrt{12}$ closer to 3 or 4?

6. Is $\sqrt{8}$ closer to 2 or 3?
7. Is $\sqrt{20}$ closer to 4 or 5?
8. Is $\sqrt{53}$ closer to 7 or 8?
9. Is $\sqrt{60}$ closer to 7 or 8?
10. Is $\sqrt{6}$ closer to 2 or 3?

5.14 Simplifying Square Roots

Square roots can sometimes be simplified even if the number under the square root is not a perfect square. One of the rules of roots is that if a and b are two positive real numbers, then it is always true that $\sqrt{a \times b} = \sqrt{a} \times \sqrt{b}$. You can use this rule to simplify square roots. Look for factors that are perfect squares.

Example 28: $\sqrt{100} = \sqrt{4 \times 25} = \sqrt{4} \times \sqrt{25} = 2 \times 5 = 10$

Example 29: $\sqrt{200} = \sqrt{100 \times 2} = 10\sqrt{2}$ This is 10 multiplied by the square root of 2.

Example 30: $\sqrt{160} = \sqrt{10 \times 16} = 4\sqrt{10}$

Simplify.

1. $\sqrt{98}$	5. $\sqrt{8}$	9. $\sqrt{54}$	13. $\sqrt{90}$
2. $\sqrt{600}$	6. $\sqrt{63}$	10. $\sqrt{40}$	14. $\sqrt{175}$
3. $\sqrt{50}$	7. $\sqrt{48}$	11. $\sqrt{72}$	15. $\sqrt{18}$
4. $\sqrt{27}$	8. $\sqrt{75}$	12. $\sqrt{80}$	16. $\sqrt{20}$

5.15 Cube Roots

Cube roots look like square roots, except that there is a "3" raised in front of the root sign:

Square root of 64: $\sqrt{64}$

Cube root of 64: $\sqrt[3]{64}$

In fact, they function very much like square roots, with one important difference. Recall asking, "What is the square root of 64?" means:

"What number multiplied by itself equals 64?"

Asking "What is the cube root of 64?" means:

"What number multiplied 3 times ('cubed') by itself equals 64?"

The answer is 4. $4 \times 4 \times 4 = 64$.

Find the cube root of the following numbers.

Examples: $\sqrt[3]{27}$ $3 \times 3 \times 3 = 27$ so $\sqrt[3]{27} = 3$

$\sqrt[3]{1000}$ $10 \times 10 \times 10 = 1000$ so $\sqrt[3]{1000} = 10$

$\sqrt[3]{\frac{8}{125}}$ $\sqrt[3]{\frac{8}{125}} = \frac{\sqrt[3]{8}}{\sqrt[3]{125}} = \frac{2 \times 2 \times 2 = 8}{5 \times 5 \times 5 = 125}$ so $\sqrt[3]{\frac{8}{125}} = \frac{2}{5}$

Find the cube roots of the following numbers.

1. $\sqrt[3]{1}$	3. $\sqrt[3]{64}$	5. $\sqrt[3]{27}$	7. $\sqrt[3]{1000}$
2. $\sqrt[3]{8}$	4. $\sqrt[3]{125}$	6. $\sqrt[3]{\frac{64}{27}}$	8. $\sqrt[3]{\frac{125}{1000}}$

Chapter 5   Exponents and Roots

Chapter 5 Review

Simplify the following problems. Remember, all answers must be written with positive exponents.

1. $4^7 \times 4^2$

2. $\left(6^3\right)^2$

3. $\left(2a^4\right)^3$

4. $10y^{-5}$

5. $\left(7b^3\right)^{-4}$

6. $y^4 \cdot y^{-3}$

7. $5^2 \cdot 5^3$

8. $\left(8^2\right)^3$

9. $\left(4^1\right)^{-5}$

10. $x^{-9} \cdot x^3$

Convert the following numbers to scientific notation.

11. $5,630,000,000$

12. $84,100,000$

13. $111,000,000,000$

14. $37,200$

15. 0.00088

16. 0.000000003654

17. 0.00000039

18. 0.00000005147

19. $3,100,000$

20. $7,892,000,000$

Use order of operations to solve. PEMDAS

21. $(8^2 + 1) \times (4^3 - 7)$

22. $(11^2 - 1) \div 10 + 3^3$

23. $8^2 + 12 \times 2 - 15$

Solve the following problems.

24. If $\dfrac{x^9}{x^7} = 49$, then what is the value of $\dfrac{x^4}{x^2}$?

25. If $\dfrac{x^{63}}{x^{59}} = 16$, then what is the value of $\dfrac{x^8}{x^4}$?

Simplify.

26. $\sqrt{88}$ 27. $\sqrt{300}$ 28. $\sqrt{49}$ 29. $\sqrt{63}$ 30. $\sqrt{8}$ 31. $\sqrt{12}$

Find the cube roots of the following numbers.

32. $\sqrt[3]{1}$ 33. $\sqrt[3]{27}$ 34. $\sqrt[3]{1,331}$ 35. $\sqrt[3]{125}$ 36. $\sqrt[3]{8}$ 37. $\sqrt[3]{\frac{1}{64}}$

Copyright © American Book Company

Chapter 5 Test

1. Rewrite the following problem using an exponent.

   $63 \times 63 \times 63 \times 63 \times 63$

   A. 63^3

   B. 63^4

   C. 63^5

   D. 63^6

   E. 6^5

2. What is the value of $x^4 \times x^5$?

   F. x^2

   G. x^4

   H. x^5

   J. x^{20}

   K. x^9

3. What is the value of $(12^2)^4$?

   A. 12^6

   B. 12^8

   C. 12^2

   D. 144^{26}

   E. 144^8

4. Which of the following is equal to x^6?

   F. $(x^2)^2$

   G. $(x^3)^2$

   H. $(x^3) \times 2$

   J. $(x^2) \times 2$

   K. $(x^4)^2$

5. Which is equivalent to the expression $\frac{1}{2}(13 - 5) + (4 + 8)^2$?

   A. 148

   B. 144

   C. 20

   D. 16^6

   E. 146

6. What is the value of $\dfrac{4x^{-5}}{7}$?

   F. $\dfrac{7x}{4x^5}$

   G. $\dfrac{4}{7x}$

   H. $\dfrac{4}{7x^5}$

   J. $\dfrac{7x^5}{4}$

   K. $\dfrac{1}{28x^5}$

7. What is the value of $7^{-2} \cdot 7^4$?

   A. $\dfrac{1}{49}$

   B. $\dfrac{7}{49}$

   C. 7

   D. 49

   E. $\dfrac{1}{7^6}$

Copyright © American Book Company

Chapter 5   Exponents and Roots

8. What is the value of $\dfrac{(6x)^{-2}}{4x^2}$?

   F. $\dfrac{1}{24x^4}$

   G. $\dfrac{1}{144x^4}$

   H. $2x^4$

   J. $2x^2$

   K. $\dfrac{1}{4x}$

9. Which is equivalent to the expression $8 \times 9 + 4^2 - (63 \div 7)$?

   A. 1143
   B. 88
   C. 78
   D. 65^6
   E. 79

10. Which number is equal to 1.7×10^5?

   F. $17,000$
   G. $170,000$
   H. $1,700,000$
   J. $17,000,000$
   K. $170,000,000$

11. Which number is equal to 9.81×10^{-3}?

   A. 981
   B. 9.81
   C. 0.981
   D. 0.0981
   E. 0.00981

12. What is $\sqrt{60}$ in simplest terms?

   F. $\sqrt{60}$
   G. $2\sqrt{15}$
   H. $5\sqrt{2}$
   J. $15\sqrt{2}$
   K. $6\sqrt{10}$

13. Which number is $\sqrt{53}$ closest to?

   A. 4
   B. 6
   C. 8
   D. 7
   E. 9

14. What is the estimated value of $\sqrt{3}$?

   F. 1.5
   G. 2
   H. 1.25
   J. 3
   K. 1.75

15. Which number below is equal to $\sqrt[3]{27}$?

   A. 4
   B. 6
   C. 3
   D. 7
   E. 2

58   Copyright © American Book Company

Chapter 6
Ratios and Proportions

6.1   Ratio Problems

In some word problems, you may be asked to express answers as a **ratio**. Ratios can look like fractions. Numbers must be written in the order they are requested. In the following example, 8 cups of sugar are mentioned before the 6 cups of strawberries. But in the question part of the example, you are asked for the ratio of STRAWBERRIES to SUGAR. The amount of strawberries IS THE FIRST WORD MENTIONED, so it must be the **top** number of the fraction. The amount of sugar, IS THE SECOND WORD MENTIONED, must be the **bottom** number of the fraction.

Example 1:   The recipe for jam requires 8 cups of sugar for every 6 cups of strawberries. What is the ratio of strawberries to sugar in this recipe?

First number requested $\dfrac{6}{8}$ cups strawberries
Second number requested   cups sugar

Answers may be reduced to lowest terms. $\dfrac{6}{8} = \dfrac{3}{4}$

This ratio could also be expressed as $3 : 4$.

Practice writing ratios for the following word problems and reduce to lowest terms. DO NOT CHANGE ANSWERS TO MIXED NUMBERS. Ratios should be left in fraction form.

1. Out of the 301 seniors, 117 are boys. What is the ratio of boys to the total number of seniors?

2. A skyscraper that stands 530 feet tall casts a shadow that is 84 feet long. What is the ratio of the shadow to the height of the skyscraper?

3. Twenty boxes of paper weigh 460 pounds. What is the ratio of boxes to pounds?

4. The newborn weighs 7 pounds and is 21 inches long. What is the ratio of weight to length?

5. Jack paid $4.00 for 6 pounds of apples. What is the ratio of the price of apples to the pounds of apples?

6. It takes 11 cups of flour to make 3 loaves of bread. What is the ratio of cups of flour to loaves of bread?

Copyright © American Book Company

59

Chapter 6   Ratios and Proportions

6.2   Solving Proportions

Two **ratios (fractions)** that are **equal** to each other are called **proportions**. For example, $\dfrac{1}{4} = \dfrac{2}{8}$. Notice that the cross products are equal: $(4)(2) = (1)(8)$. **Read the following example to see how to find a number missing from a proportion.**

Example 2: $\dfrac{5}{15} = \dfrac{8}{x}$

Step 1:   To find x, you first multiply the two numbers that are diagonal to each other.

$$\frac{5}{\{15\}} = \frac{\{8\}}{x}$$

$$15 \times 8 = 120$$

$$5 \times x = 5x$$

Therefore, $5x = 120$.

Step 2:   Then solve for x by dividing the product (120) by the other number in the proportion (5).

$$120 \div 5 = 24$$

Therefore, $\dfrac{5}{15} = \dfrac{8}{24}$ and $x = 24$.

Practice finding the number missing from the following proportions. First, multiply the two numbers that are diagonal from each other. Then solve for x by dividing by the other number.

1. $\dfrac{2}{4} = \dfrac{9}{x}$

2. $\dfrac{9}{3} = \dfrac{x}{7}$

3. $\dfrac{x}{12} = \dfrac{3}{6}$

4. $\dfrac{2}{x} = \dfrac{4}{12}$

5. $\dfrac{15}{x} = \dfrac{5}{3}$

6. $\dfrac{8}{x} = \dfrac{2}{5}$

7. $\dfrac{14}{6} = \dfrac{x}{3}$

8. $\dfrac{1}{x} = \dfrac{8}{64}$

9. $\dfrac{8}{2} = \dfrac{x}{3}$

10. $\dfrac{16}{2} = \dfrac{x}{4}$

11. $\dfrac{5}{6} = \dfrac{35}{x}$

12. $\dfrac{2}{x} = \dfrac{3}{18}$

13. $\dfrac{x}{4} = \dfrac{4}{16}$

14. $\dfrac{2}{5} = \dfrac{x}{40}$

15. $\dfrac{8}{4} = \dfrac{16}{x}$

16. $\dfrac{x}{2} = \dfrac{7}{14}$

17. $\dfrac{6}{12} = \dfrac{x}{8}$

18. $\dfrac{x}{40} = \dfrac{5}{20}$

19. $\dfrac{4}{8} = \dfrac{x}{4}$

20. $\dfrac{1}{4} = \dfrac{42}{x}$

21. $\dfrac{x}{32} = \dfrac{1}{2}$

6.3 Ratio and Proportion Word Problems

You can use ratios and proportions to solve problems.

Example 3: A stick one meter long is held perpendicular to the ground and casts a shadow 0.4 meters long. At the same time, an electrical tower casts a shadow 112 meters long. Use ratio and proportion to find the height of the tower.

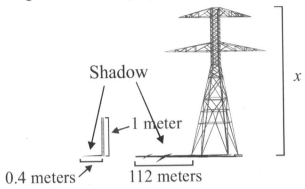

Step 1: Set up a proportion using the numbers in the problem. Put the shadow lengths on one side of the equation and put the heights on the other side. The 1 meter height is paired with the 0.4 meter length, so let them both be top numbers. Let the unknown height be x.

$$\begin{array}{cc} \textbf{shadow} & \textbf{object} \\ \textbf{length} & \textbf{height} \end{array}$$

$$\frac{0.4}{112} = \frac{1}{x}$$

Step 2: Solve the proportion as you did on page 60.

$$112 \times 1 = 112 \qquad 112 \div 0.4 = 280$$

Answer: The tower height is 280 meters.

Use ratios and proportions to solve the following problems.

1. If 4 pounds of jelly beans cost $6.82, how much would 2 pounds cost?

2. Ashley drove from Memphis, Tennessee to Atlanta, Georgia. She drove for 5 hours and 45 minutes and drove 391 miles. How far did she travel in one hour?

3. Out of every 6 students surveyed, 1 listens to country music. At that rate, how many students in a school of 1,200 listen to country music?

4. Bailey, a Labrador retriever, has a litter of 10 puppies. Four are black. At that rate, how many puppies in a litter of 5 would be black?

5. According to the instructions on a bag of fertilizer, 5 pounds of fertilizer are needed for every 50 square feet of lawn. How many square feet will a 15-pound bag cover?

Chapter 6   Ratios and Proportions

6. Faye wants to know how tall her school building is. On a sunny day, she measures the shadow of the building to be 8 feet. At the same time she measures the shadow cast by a 4 foot statue to be 2 feet. How tall is her school building?

7. For the first 3 home football games, the concession stand sold a total of 600 hotdogs. If that ratio stays constant, how many hotdogs will sell for all 8 home games?

8. A race car can travel 2 laps in 4 minutes. At this rate, how long will it take the race car to complete 250 laps?

9. If it takes 8 cups of flour to make 4 loaves of bread, how many loaves of bread can you make from 40 cups of flour?

10. Rudolph can mow a lawn that measures $1,000$ square feet in 3 hours. At that rate, how long would it take him to mow a lawn $4,500$ square feet?

11. Kathryn goes 12 mph when she rides her bike. How far does she go in 25 minutes?

12. Danny can paint $4\frac{1}{2}$ rooms in 3 hours. How many rooms can he paint in 5 hours?

13. Casey can do 4 math problems in $1\frac{3}{4}$ minutes. How many can he do in 7 minutes?

14. Sarah reads $1\frac{1}{4}$ pages in $1\frac{1}{2}$ minutes. How many pages does she read in 2 minutes?

15. Cameron runs $1\frac{1}{3}$ miles in 8 minutes. How many miles can he run in 20 minutes?

16. Jessie can swim 75 meters in $1\frac{3}{4}$ minutes. How far can he swim in 5 minutes?

6.4 Maps and Scale Drawings

Example 4: On a map drawn to scale, 5 cm represents 30 kilometers. A line segment connecting two cities is 7 cm long. What distance does this line segment represent?

Step 1: Set up a proportion using the numbers in the problem. Keep centimeters on one side of the equation and kilometers on the other. The 5 cm is paired with the 30 kilometers, so let them both be top numbers. Let the unknown distance be x.

$$\begin{array}{cc} \textbf{cm} & \textbf{km} \\ \dfrac{5}{7} & = \dfrac{30}{x} \end{array}$$

Step 2: Solve the proportion as you have previously.
$$7 \times 30 = 210$$
$$210 \div 5 = 42$$

Answer: 7 cm represents 42 km.

Sometimes the answer to a scale drawing problem will be a fraction or a mixed number.

Example 5: On a scale drawing, 2 inches represents 30 feet. How many inches long is a line segment that represents 5 feet?

Step 1: Set up the proportion as you did above.

$$\begin{array}{cc} \textbf{inches} & \textbf{feet} \\ \dfrac{2}{x} & = \dfrac{30}{5} \end{array}$$

Step 2: First multiply the two numbers that are diagonal from each other. Then divide by the other number.
$$2 \times 5 = 10$$
$10 \div 30$ is less than 1 so express the answer as a fraction and simplify.
$$10 \div 30 = \frac{10}{30} = \frac{1}{3} \text{ inch}$$

Set up proportions for each of the following problems and solve.

1. If 3 inches represents 50 miles on a scale drawing, how long would a line segment be that represents 200 miles?

2. On a scale drawing, 2 cm represents 10 km. A line segment on the drawing is 5 cm long. What distance does this line segment represent?

3. On a map drawn to scale, 4 cm represents 280 km. How many kilometers are represented by a line 5 cm long?

4. If 4 inches represents 80 miles on a scale drawing, how long would a line segment be that represents 240 miles?

5. On a map drawn to scale, 5 cm represents 100 km. How long would a line segment be that represents 160 km?

6. On a scale drawing of a house plan, one inch represents 10 feet. How many feet wide is the bathroom if the width on the drawing is 2.5 inches?

Copyright © American Book Company

Chapter 6   Ratios and Proportions

Chapter 6 Review

Answer the following ratio and proportion questions.

1. Out of 250 coins, 50 are in mint condition. What is the ratio of mint condition coins to the total number of coins?

2. The ratio of boys to girls in the ninth grade is 6 : 5. If there are 185 girls in the class, how many boys are there?

3. Forty-six out of the total 310 seniors graduate with honors. What is the ratio of seniors graduating with honors to the total number of seniors?

4. Aunt Bess uses 3 cups of oatmeal to bake 6-dozen oatmeal cookies. How many cups of oatmeal would she need to bake 18-dozen cookies?

5. On a map, 3 centimeters represents 150 kilometers. If a line between two cities measures 4.5 centimeters, how many kilometers apart are they?

6. Shondra used four ounces of chocolate chips to make two dozen cookies. At that rate, how many ounces of chocolate chips would she need to make five dozen cookies?

7. When Rick measures the shadow of a yard stick, it is 7 inches. At the same time, the shadow of the tree he would like to chop down is 42 inches. How tall is the tree in yards?

8. The ratio of boys to girls in the gym class at a local middle school is 3 : 2. What percent of the gym class is girls?

9. There is a bag of marbles that has 40% blue marbles, 10% red marbles, 30% green marbles, and 20% black marbles. What is the ratio of red marbles to black marbles?

10. Johnny is looking at a map and measures the distance between Chattanooga and Nashville to be 3.5 inches. He wants to know the actual distance between the two cities. According to the scale on the map, 1 inch = 38 miles. How many miles is there between Chattanooga and Nashville?

11. Gary Norbert runs 4 miles then walks 1 mile. How many miles does he run in every 10 miles?

Solve the following proportions.

12. $\dfrac{4}{x} = \dfrac{1}{3}$

13. $\dfrac{3}{6} = \dfrac{x}{10}$

14. $\dfrac{x}{6} = \dfrac{4}{3}$

15. $\dfrac{2}{9} = \dfrac{6}{x}$

16. $\dfrac{9}{x} = \dfrac{3}{5}$

17. $\dfrac{15}{x} = \dfrac{1}{5}$

18. $\dfrac{7}{17} = \dfrac{x}{850}$

19. $\dfrac{x}{15} = \dfrac{5}{3}$

20. $\dfrac{3}{8} = \dfrac{x}{24}$

64   Copyright © American Book Company

Chapter 6 Test

1. Gorden collects rocks. He has 35 sedimentary rocks, 25 igneous rocks, 15 metamorphic rocks, and 10 unknown rocks. Which ratio compares Gorden's sedimentary rocks to his unknown rocks, expressed in simplest form?

   A. $\dfrac{35}{10}$

   B. $35 : 10$

   C. $7 : 2$

   D. $2 : 7$

   E. $35 : 85$

2. Applewood Elementary boasts a student to teacher ratio of $14 : 1$. What percent of the school makes up the number of teachers at Applewood Elementary? (Round percent to the nearest tenth.)

   F. 6.7%

   G. 59%

   H. 82.9%

   J. 4.5%

   K. 15%

3. Find n: $\dfrac{3}{4} = \dfrac{75}{n}$.

   A. $n = 1$

   B. $n = 10$

   C. $n = 50$

   D. $n = 25$

   E. $n = 100$

4. The ratio of a rectangle's length to its width is $2 : 1$. If the length is 14 cm, what is its width?

   F. 14 cm

   G. 10 cm

   H. 1 cm

   J. 28 cm

   K. 7 cm

5. A bag of marbles has 24 marbles. 37.5% are red marbles, 41.7% are green marbles, 12.5% are yellow marbles, and 8.3% are blue marbles. What is the ratio of the total number of marbles to yellow marbles? Express your answer in simplest terms.

   A. $\dfrac{24}{21}$

   B. $8 : 1$

   C. $8 : 7$

   D. $1 : 8$

   E. $24 : 12.5$

6. Solve for x: $\dfrac{11}{9} = \dfrac{121}{x}$

   F. 11

   G. 12

   H. 0.81

   J. 99

   K. 90

7. An air tank that is 500 mL is 80% oxygen and 20% nitrogen. What is the amount of oxygen in milliliters in a 200 mL air tank that contains the same ratio?

   A. 160 mL

   B. 40 mL

   C. 400 mL

   D. 100 mL

   E. 150 mL

8. Chris bought six pounds of ground beef for $18.60. How much would 15 pounds of ground beef cost at the same price per pound?

   F. $18.60

   G. $65.80

   H. $37.20

   J. $279.00

   K. $46.50

Copyright © American Book Company

Chapter 7
Introduction to Graphing

7.1 Graphing on a Number Line

Number lines allow you to graph values of positive and negative numbers as well as zero. Any real number, whether it is a fraction, decimal, or integer, can be plotted on a number line. Number lines can be horizontal or vertical. The examples below illustrate how to plot different types of numbers on a number line.

7.2 Graphing Fractional Values

Example 1: What number does point A represent on the number line below?

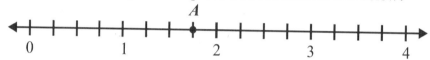

Step 1: Point A is between the numbers 1 and 2, so it is greater than 1 but less than 2. We can express the value of A as a fractional value that falls between 1 and 2. To do so, copy the integer that point A falls between which is closer to zero on the number line. In this case, copy the 1 because 1 is closer to zero on the number line than the 2.

Step 2: Count the number of spaces between each integer. In this case, there are 4 spaces between the 1 and the 2. Put this number as the bottom number in your fraction.

Step 3: Count the number of spaces between the 1 and the point A. Point A is 3 spaces away from number 1. Put this number as the top number in your fraction.

Point A is at $1\frac{3}{4}$ ← The integer that point A falls between that is closest to 0
← The number of spaces between 1 and A
← The number of spaces between 1 and 2

* Always want to go toward 0.

7.2 Graphing Fractional Values

Example 2: What number does point B represent on the number line below?

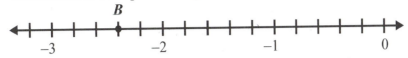

Step 1: Point B is between -2 and -3. Again, we can express the value of B as a fraction that falls between -2 and -3. Copy the integer that point B falls between which is closer to zero. The -2 is closer to zero than -3, so copy -2.

Step 2: In this example, there are 5 spaces between each integer. Five will be the bottom number in the fraction.

Step 3: There are 2 spaces between -2 and point B. Two will be the top number in the fraction.
Point B is at $-2\frac{2}{5}$.

Determine and record the value of each point on the number lines below.

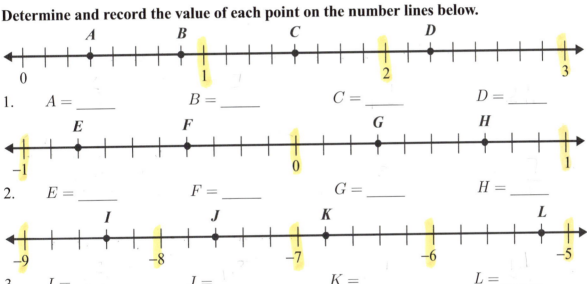

1. $A = $ ____ $B = $ ____ $C = $ ____ $D = $ ____

2. $E = $ ____ $F = $ ____ $G = $ ____ $H = $ ____

3. $I = $ ____ $J = $ ____ $K = $ ____ $L = $ ____

4. $M = $ ____ $N = $ ____ $P = $ ____ $Q = $ ____

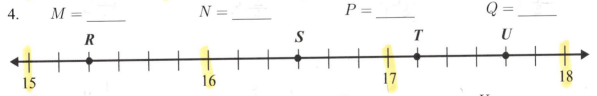

5. $R = $ ____ $S = $ ____ $T = $ ____ $U = $ ____

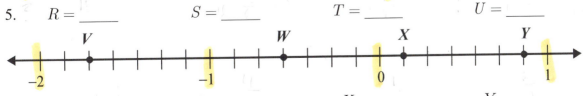

6. $V = $ ____ $W = $ ____ $X = $ ____ $Y = $ ____

Chapter 7 Introduction to Graphing

7.3 Plotting Points on a Vertical Number Line

Number lines can also be drawn up and down **(vertical)** instead of across the page **(horizontal)**. You plot points on a vertical number line the same way as you do on a horizontal number line.

Record the value represented by each point on the number lines below.

1. $A = $ ____
2. $B = $ ____
3. $C = $ ____
4. $D = $ ____
5. $E = $ ____
6. $F = $ ____
7. $G = $ ____
8. $H = $ ____

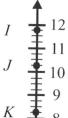

17. $Q = $ ____
18. $R = $ ____
19. $S = $ ____
20. $T = $ ____
21. $U = $ ____
22. $V = $ ____
23. $W = $ ____
24. $X = $ ____

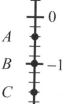

9. $I = $ ____
10. $J = $ ____
11. $K = $ ____
12. $L = $ ____
13. $M = $ ____
14. $N = $ ____
15. $P = $ ____
16. $Q = $ ____

25. $A = $ ____
26. $B = $ ____
27. $C = $ ____
28. $D = $ ____
29. $E = $ ____
30. $F = $ ____
31. $G = $ ____
32. $H = $ ____

7.4 Graphing Rational Numbers on a Number Line

Improper fractions, decimal values, and all other rational numbers can be plotted on a number line. Study the examples below.

Example 3: Where would $\frac{4}{3}$ fall on the number line below?

Step 1: Convert the improper fraction to a mixed number. $\frac{4}{3} = 1\frac{1}{3}$

Step 2: $1\frac{1}{3}$ is $\frac{1}{3}$ of the distance between the numbers 1 and 2. Estimate this distance by dividing the distance between points 1 and 2 into thirds. Plot the point at the first division.

Example 4: Plot the value of -1.75 on the number line below.

Step 1: Convert the value -1.75 to a mixed fraction. $-1.75 = -1\frac{3}{4}$

Step 2: $-1\frac{3}{4}$ is $\frac{3}{4}$ of the distance between the numbers -1 and -2. Estimate this distance by dividing the distance between points -1 and -2 into fourths. Plot the point at the third division.

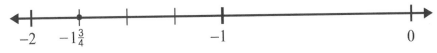

Example 5: Plot the value of $3.5 \div 2$ on the number line below.

Step 1: Figure the value of $3.5 \div 2$. $3.5 \div 2 = 1.75$.

Step 2: Plot 1.75.

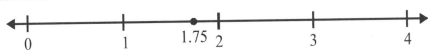

Chapter 7 Introduction to Graphing

Plot and label the following values on the number lines given below.

1. $A = \dfrac{5}{4}$ $B = \dfrac{12}{5} = 2\dfrac{2}{5}$ $C = \dfrac{2}{3}$ $D = -\dfrac{3}{2}$

2. $E = 1.4$ $F = -2.25$ $G = -0.6$ $H = 0.625$

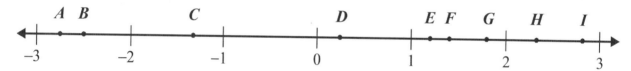

3. $I = 0.25$ $J = 0.9$ $K = 1.9$ $L = 2.6$

Match the correct value for each point on the number line below.

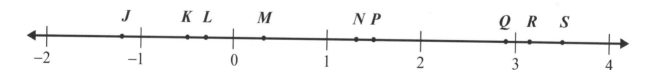

4. $1.8 = $ ___
5. $\dfrac{7}{3} = $ ___
6. $1\dfrac{1}{3} = $ ___
7. $-\dfrac{5}{2} = $ ___
8. $-2.75 = $ ___
9. $-\dfrac{4}{3} = $ ___
10. $2\dfrac{4}{5} = $ ___
11. $\dfrac{6}{5} = $ ___
12. $0.25 = $ ___

13. $\dfrac{19}{6} = $ ___
14. $-0.5 = $ ___
15. $\dfrac{5}{4} = $ ___
16. $\dfrac{1}{3} = $ ___
17. $1.5 = $ ___
18. $-0.3 = $ ___
19. $-\dfrac{6}{5} = $ ___
20. $3\dfrac{1}{2} = $ ___
21. $2.9 = $ ___

7.5 Rectangle Coordinate System

A **Cartesian coordinate plane** allows you to graph points with two values. A Cartesian coordinate plane is made up of two number lines. The horizontal number line is called the x-**axis**, and the vertical number line is called the y-**axis**. The point where the x and y axes intersect is called the **origin**. The x and y axes separate the Cartesian coordinate plane into four quadrants that are labeled I, II, III, and IV. The quadrants are labeled and explained on the graph below. Each point graphed on the plane is designated by an **ordered pair** of coordinates. For example, $(2, -1)$ is an ordered pair of coordinates designated by **point B** on the plane below. The first number, 2, tells you to go over positive two on the x-axis. The -1 tells you to then go down negative one on the y-axis.

Remember: The first number always tells you how far to go right or left of 0, and the second number always tells you how far to go up or down from 0.

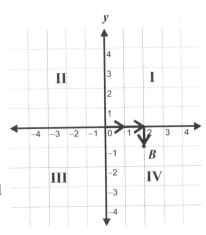

Quadrant II:
The x-coordinate is negative, and the y-coordinate is positive $(-,+)$.

Quadrant III:
Both coordinates in the ordered pair are negative $(-,-)$.

Quadrant I:
Both coordinates in the ordered pair are positive $(+,+)$.

Quadrant IV:
The x-coordinate is positive and the y-coordinate is negative $(+,-)$.

Plot and label the following points on the Cartesian coordinate plane provided.

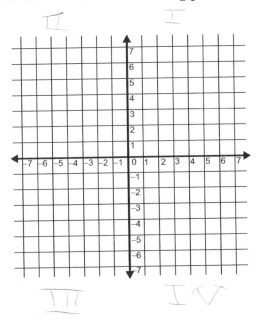

A. $(3, 5)$
B. $(2, -1)$
C. $(-3, 4)$
D. $(2, -2)$
E. $(4, 1)$
F. $(0, 3)$

G. $(3, 3)$
H. $(-2, 4)$
I. $(4, -5)$
J. $(4, -3)$
K. $(2, 2)$
L. $(5, -4)$

M. $(0, 0)$
N. $(-4, -2)$
O. $(-2, -2)$
P. $(-4, 2)$
Q. $(-3, -4)$
R. $(-2, 1)$

Chapter 7 Introduction to Graphing

7.6 Ordered Pairs

When identifying **ordered pairs**, count how far left or right of 0 to find the x-coordinate and then how far up or down from 0 to find the y-coordinate.

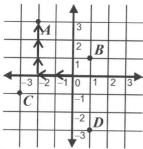

Point A: Left (negative) two and up (positive) three $= (-2, 3)$ in quadrant II

Point B: Right (positive) one and up (positive) one $= (1,1)$ in quadrant I

Point C: Left (negative) three and down (negative) one $= (-3, -1)$ in quadrant III

Point D: Right (positive one and down (negative) three $= (1, -3)$ in quadrant IV

Fill in the ordered pair for each point, and tell which quadrant it is in.

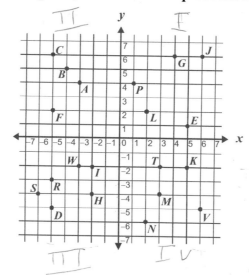

1. point A = (,) quadrant____
2. point B = (,) quadrant____
3. point C = (,) quadrant____
4. point D = (,) quadrant____
5. point E = (,) quadrant____
6. point F = (,) quadrant____
7. point G = (,) quadrant____
8. point H = (,) quadrant____
9. point I = (,) quadrant____
10. point J = (,) quadrant____

11. point K = (,) quadrant____
12. point L = (,) quadrant____
13. point M = (,) quadrant____
14. point N = (,) quadrant____
15. point P = (,) quadrant____
16. point R = (,) quadrant____
17. point S = (,) quadrant____
18. point T = (,) quadrant____
19. point V = (,) quadrant____
20. point W = (,) quadrant____

7.6 Ordered Pairs

Sometimes, points on a coordinate plane fall on the x or y axis. If a point falls on the x-axis, then the second number of the ordered pair is 0. If a point falls on the y-axis, the first number of the ordered pair is 0.

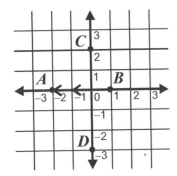

Point A: Left (negative) two and up zero $= (-2, 0)$
Point B: Right (positive) one and up zero $= (1, 0)$
Point C: Left/right zero and up (positive) two $= (0, 2)$
Point D: Left/right zero and down (negative) three $= (0, -3)$

Fill in the ordered pair for each point.

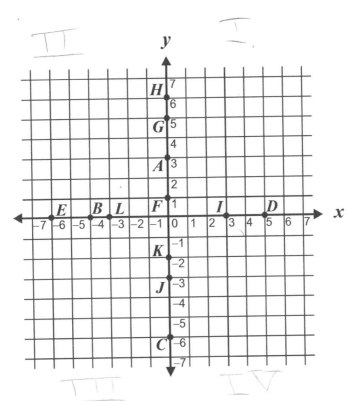

1. point $A = (\ ,\ )$
2. point $B = (\ ,\ )$
3. point $C = (\ ,\ )$
4. point $D = (\ ,\ )$
5. point $E = (\ ,\ )$
6. point $F = (\ ,\ )$
7. point $G = (\ ,\ )$
8. point $H = (\ ,\ )$
9. point $I = (\ ,\ )$
10. point $J = (\ ,\ )$
11. point $K = (\ ,\ )$
12. point $L = (\ ,\ )$

Copyright © American Book Company

Chapter 7 Review

1.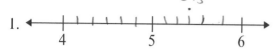
   Plot and label $5\frac{3}{5}$ on the number line above.

2.
   Plot and label $-3\frac{1}{2}$ on the number line above.

3. ←—+—+—+—+—+—→
      6 7 8 9 10
   Plot and label 7.2 on the number line above.

4. ←—+——+——+—→
     −3   −2   −1
   Plot and label −2.3 on the number line above.

Record the value represented by the point on the number line for questions 5–10.

5. $A = $ _____
6. $B = $ _____
7. $C = $ _____
8. $D = $ _____
9. $E = $ _____
10. $F = $ _____

Record the coordinates and quadrants of the following points.

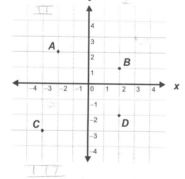

	Coordinates	Quadrants
11. $A =$	_____	_____
12. $B =$	_____	_____
13. $C =$	_____	_____
14. $D =$	_____	_____

On the same plane above, label these additional coordinates.

15. $E = (2, -4)$
16. $F = (3, 3)$
17. $G = (-3, 1)$
18. $H = (-2, -3)$

Answer the following questions.

19. In which quadrant does the point $(4, -3)$ lie?

20. In which quadrant does the point $(-2, 1)$ lie?

Chapter 7 Test

1. Which point on the number line represents -1.2?

   A. P
   B. Q
   C. R
   D. S
   E. none of the above

2. Which point on the number line represents $-2\frac{3}{4}$?

   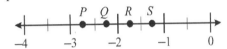

   F. P
   G. Q
   H. R
   J. S
   K. none of the above

3. What are the coordinates of point U on the grid below?

   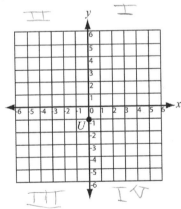

   A. $(0, 1)$
   B. $(0, -1)$
   C. $(-1, 0)$
   D. $(1, 0)$
   E. $(-1, -1)$

4. Which of these is the best estimate of the coordinate of Point P on the number line?

   F. $-1\frac{1}{8}$
   G. $-1\frac{3}{8}$
   H. $-1\frac{5}{8}$
   J. $-2\frac{3}{8}$
   K. $1\frac{5}{8}$

5. Which point on the number line represents $-\frac{12}{10}$?

   A. none of the above
   B. P
   C. R
   D. S
   E. Q

6. Which point is located at $(3, 3)$?

   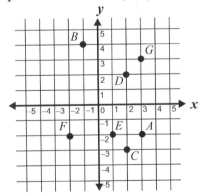

   F. D
   G. E
   H. F
   J. G
   K. A

Chapter 8
Introduction to Algebra

8.1   Algebra Vocabulary

Vocabulary Word	Example	Definition
variable	$4x$ (x is the variable)	a letter that can be replaced by a number
coefficient	$4x$ (4 is the coefficient)	a number multiplied by a variable or variables
term	$5x^2 + x - 2$ ($5x^2$, x, and -2 are terms)	numbers or variables separated by $+$ or $-$ signs
constant	$5x + 2y + 4$ (4 is a constant)	a term that does not have a variable
degree	$4x^2 + 3x - 2$ (the degree is 2)	the largest power of a variable in an expression
leading coefficient	$4x^2 + 3x - 2$ (4 is the leading coefficient)	the number multiplied by the term with the highest power
sentence	$2x = 7$ or $5 \leq x$	two algebraic expressions connected by $=, \neq, <, >, \leq, \geq,$ or $\approx$
equation	$4x = 8$	a sentence with an equal sign
inequality	$7x < 30$ or $x \neq 6$	a sentence with one of the following signs: $\neq, <, >, \leq, \geq,$ or $\approx$
base	6^3 (6 is the base)	the number used as a factor
exponent	6^3 (3 is the exponent)	the number of times the base is multiplied by itself

76   Copyright © American Book Company

8.2 Substituting Numbers for Variables

These problems may look difficult at first glance, but they are very easy. Simply replace the variable with the number the variable is equal to, and solve the problems.

Example 1: In the following problems, substitute 3 for b.

Problem	Calculation	Solution
1. $b + 2$	Simply replace the b with 3. $3 + 2$	5
2. $13 - b$	$13 - 3$	10
3. $4b$	This means multiply. 4×3	12
4. $\dfrac{27}{b}$	This means divide. $27 \div 3$	9
5. b^3	$3 \times 3 \times 3$	27
6. $4b + 8$	$(4 \times 3) + 8$	20

Note: Be sure to do all multiplying and dividing before adding and subtracting.

Example 2: In the following problems, let $a = 3$, $b = -1$, and $c = 4$.

Problem	Calculation	Solution
1. $3ac - b$	$3 \times 3 \times 4 - (-1)$	37
2. $cb^2 + 2$	$4 \times (-1)^2 + 2 = 4 \times 1 + 2$	6
3. $\dfrac{ac}{3}$	$(3 \times 4) \div 3 = 12 \div 3$	4

In the following problems, r $= 9$. Solve the problems.

1. $r - 3 =$

2. $11 + r =$

3. $\dfrac{63}{r} =$

4. $r^2 + 1 =$

5. $4r - 6 =$

6. $r^2 - 80 =$

7. $\dfrac{r^2}{9} =$

8. $3r + 4 =$

9. $6r \div 2 =$

In the following problems r $= -4$, s $= 10$, t $= 2$, and u $= -1$. Solve the problems.

10. $-\frac{1}{4}r + \frac{1}{8}t =$

11. $\dfrac{tu}{9} =$

12. $s - 7 =$

13. $2ut + r =$

14. $\dfrac{3t}{2} =$

15. $u^2 + s + 1 =$

16. $3(4 + t) =$

17. $s - 9 + t =$

18. $\frac{3}{2}r^2 + 4 =$

19. $rtu =$

20. $3s + \frac{1}{5}sr =$

21. $rs + tu =$

Copyright © American Book Company

Chapter 8   Introduction to Algebra

8.3   Understanding Algebra Word Problems

The biggest challenge to solving word problems is figuring out whether to add, subtract, multiply, or divide. Below is a list of key words and their meanings. This list does not include every situation you might see, but it includes the most common examples.

Words Indicating Addition	**Example**	**Add**
and	3 **and** 9	$3 + 9$
increased	The original price of $14 **increased** by $2.	$14 + 2$
more	7 coins and 3 **more**	$7 + 3$
more than	Josh has 15 points. Will has 3 **more than** Josh.	$15 + 3$
plus	2 baseballs **plus** 1 baseballs	$2 + 1$
sum	the **sum** of 4 and 2	$4 + 2$
total	the **total** of 9, 5, and 11	$9 + 5 + 11$

Words Indicating Subtraction	**Example**	**Subtract**
decreased	$19 **decreased** by $7	$19 - 7$
difference	the **difference** between 24 and 10	$24 - 10$
less	12 days **less** 5	$12 - 5$
less than	Jose completed 11 laps **less than** Mike's 15.	$*15 - 11$
left	Ray sold 22 out of 40 tickets. How many did he have **left**?	$*40 - 22$
lower than	This month's rainfall is 3 inches **lower than** last month's rainfall of 9 inches.	$*9 - 3$
minus	8 **minus** 7	$8 - 7$

* In subtraction word problems, you cannot always subtract the numbers in the order that they appear in the problem. Sometimes the first number should be subtracted from the last. You must read each problem carefully.

Words Indicating Multiplication	**Example**	**Multiply**
triple	Her $150 profit **tripled** in in a month.	150×3
half	**Half** of the $800 collected went to charity.	$\frac{1}{2} \times 800$
product	the **product** of 5 and 11	5×11
times	Li scored 5 **times** as many points as Ted who only scored 3.	5×3
double	The bacteria **doubled** its original colony of 5,000 in just one day.	$2 \times 5,000$
twice	Ron has 8 CDs. Tom has **twice** as many.	2×8

Words Indicating Division	**Example**	**Divide**
divide into, by, or among	The group of 20 **divided into** 5 teams	$20 \div 5$ or $\frac{20}{5}$
quotient	the **quotient** of 36 and 4	$36 \div 4$ or $\frac{36}{4}$

78   Copyright © American Book Company

8.3   Understanding Algebra Word Problems

Match the phrase with the correct algebraic expression below. The answers will be used more than once.

A. $x + 4$

B. $4x$

C. $4 - x$

D. $x - 4$

E. $\dfrac{x}{4}$

1. 4 more than x

2. x divided into 4

3. 4 less than x

4. four times x

5. the quotient of x and 4

6. x increased by 4

7. 4 less x

8. the product of 4 and x

9. x decreased by 4

10. x times 4

11. 4 minus x

12. the total of 4 and x

Now practice writing parts of algebraic expressions from the following word problems.

Example 3:   the product of 3 and a number, x Answer: $3x$

13. the sum of 3 and y $3+y$

14. x minus 2 $x-2$

15. the quotient of r divided by 7 $\frac{r}{7}$

16. 5 more than p $p+5$

17. 2 less than y $2-y$

18. triple n $3n$

19. the total of h and 14 $h+14$

20. 7 less r $7-r$

21. double y $2y$

22. 2 increased by c $2+c$

23. 8 less than z $8-z$

24. half of r $\frac{r}{2}$, $\frac{1}{2}r$, $r \div 2$

25. 4 times t $4t$

26. z minus 5 $z-5$

27. 8 plus m $8+m$

28. 3 divided by s $3 \div s$

29. the product of 4 and n $4n$

30. z decreased by 10 $10-z$

31. four times as much as x $4x$

32. q less than 12 $q-12$

Copyright © American Book Company

79

Chapter 8   Introduction to Algebra

If a word problem contains the word "sum" or "difference," put the numbers that "sum" or "difference" refer to in parentheses to be added or subtracted first. Do not separate them. Look at the examples below.

Examples:

	RIGHT	WRONG
sum of 2 and 4, times 5	$5(2+4)=30$	$2+4\times 5=22$
the sum of 4 and 6, divided by 2	$\dfrac{(4+6)}{2}=5$	$4+\dfrac{6}{2}=7$
4 times the difference between 10 and 5	$4(10-5)=20$	$4\times 10-5=35$
20 divided by the difference between 4 and 2	$\dfrac{20}{(4-2)}=10$	$20\div 4-2=3$
the sum of x and 4, multiplied by 2	$2(x+4)=2x+8$	$x+4\times 2=x+8$

Change the following phrases into algebraic expressions.

1. 4 times the sum of x and 2

2. the difference between 8 and 4, divided by 2

3. 60 divided by the sum of 5 and 2

4. twice the sum of 15 and x

5. the difference between x and 7, divided by 3

6. 6 times the difference between x and 3

7. 10 multiplied by the sum of 4 and 5

8. the difference between x and 3, divided by 5

9. x divided by the sum of 7 and 2

$$\frac{\times}{(7+2)}$$

10. x minus 3, times 7

$$7(x-3)$$

11. 70 multiplied by the sum of x and 4

$$70(x+4)$$

12. twice the difference between 4 and x

$$2(4-x)$$

13. 8 times the sum of 2 and 9

$$8(2+9)$$

14. 3 times the difference between 8 and 1

$$3(8-1)$$

15. 14 divided by the sum of 3 and 11

$$\frac{14}{(3+11)}$$

16. four minus x, multiplied by 15

$$15(4-x)$$

80   Copyright © American Book Company

8.4 Setting Up Algebra Word Problems

To complete an algebra problem, an equal sign must be added. The words "is" or "are" as well as "equal(s)" signal that you should add an equal sign.

Example 4: Double Jake's age, x, minus 4 is 22.

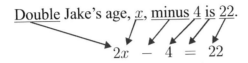

$$2x - 4 = 22$$

Translate the following word problems into algebra problems. DO NOT find the solutions to the problems yet.

1. Triple the original number, n, is 3,500.
   $3n = 3,5000$

2. The product of a number, y, and 3 is equal to 15.
   $3y = 15$

3. Four times the difference of a number, x, and 10 is 35.
   $4x = 35$

4. The total, t, divided into 7 groups is 20.
   $\frac{t}{7} = 20$

5. The number of parts in inventory, p, minus 80 parts sold today is 270.
   $p - 80 = 270$

6. One-half an amount, x, added to $45 is $342
   $\frac{1}{2}x + 45 = 342$

7. One hundred seeds divided by 4 rows equals n number of seeds per row.
   $100 \div 4 = n$

8. A number, y, less than 20 is 47.
   $y - 20 = 47$

9. His base pay of $500 increased by his commission, x, is $640.
   $500 + n = 640$

10. Seventeen more than half a number, h, is 100.
    $17 - h = 100$

11. This month's sales of $2,900 are double January's sales, x.
    $2x = \$2,900$

12. The quotient of a number, w, and 8 is 24.
    $w \div 8 = 24$

13. Six less a number, d, is 32.
    $6 - d = 32$

14. Four times the sum of a number, y, and 7 is 84.

15. We started with x number of students. When 2 moved away, we had 28 left.
    $x - 2 = 28$

16. A number, b, divided by 29 is 3.
    $b \div 29 = 3$

Chapter 8   Introduction to Algebra

8.5   Changing Algebra Word Problems to Algebraic Equations

Example 5:   There are 3 people who have a total weight of 595 pounds. Sally weighs 20 pounds less than Jessie. Rafael weighs 15 pounds more than Jessie. How much does Jessie weigh?

Step 1:   Notice everyone's weight is given in terms of Jessie. Sally weighs 20 pounds less than Jessie. Rafael weighs 15 pounds more than Jessie. First, we write everyone's weight in terms of Jessie, j.

$$\text{Jessie} = j$$
$$\text{Sally} = j - 20$$
$$\text{Rafael} = j + 15$$

Step 2:   We know that all three together weigh 595 pounds. We write the sum of everyone's weight equal to 595.

$$j + j - 20 + j + 15 = 595$$

We will learn to solve these problems in chapter 9.

Change the following word problems to algebraic equations.

1. Fluffy, Spot, and Shampy have a combined age in dog years of 82. Spot is 14 years younger than Fluffy. Shampy is 6 years older than Fluffy. What is Fluffy's age, f, in dog years?

2. Jerry Marcosi puts 8% of the amount he makes per week into a retirement account, r. He is paid $12.00 per hour and works 40 hours per week for a certain number of weeks, w. Write an equation to help him find out how much he puts into his retirement account.

3. A furniture store advertises a 35% off liquidation sale on all items. What would the sale price (p) be on a $2,742 dining room set?

4. Kyle Thornton buys an item which normally sells for a certain price, x. Today the item is selling for 25% off the regular price. A sales tax of 8% is added to the equation to find the final price, f.

5. Tamika Francois runs a floral shop. On Tuesday, Tamika sold a total of $800 worth of flowers. The flowers cost her $75, and she paid an employee to work 8 hours for a given wage, w. Write an equation to help Tamika find her profit, p, on Tuesday.

6. Sharice is a waitress at a local restaurant. She makes an hourly wage of $2.70, plus she receives tips. On Monday, she works 8 hours and receives tip money, t. Write an equation showing what Sharice makes on Monday, y.

7. Jenelle buys x shares of stock in a company at $27.80 per share. She later sells the shares at $41.29 per share. Write an equation to show how much money, m, Jenelle has made.

82   Copyright © American Book Company

8.6 Substituting Numbers in Formulas

Example 6: Area of a parallelogram: $A = b \times h$
Find the area of the parallelogram if $b = 20$ cm and $h = 10$ cm.

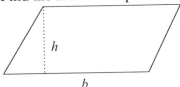

Step 1: Copy the formula with the numbers given in place of the letter in the formula.
$A = 20 \times 10$

Step 2: Solve the problem. $A = 20 \times 10 = 200$. Therefore, $A = 200$ cm^2.

Solve the following problems using the formulas given.

1. The volume of a rectangular pyramid is determined by using the following formula:
$$V = \frac{lwh}{3}$$
Find the volume of the pyramid if $l = 6$ in, $w = 6$ in, and $h = 11$ in.

2. Find the volume of a cone with a radius of 30 inches and a height of 60 inches using the formula:
$V = \frac{1}{3}\pi r^2 h$ $\pi \approx 3.14$

3. Lumber is measured by the following formula:
Number of board feet $= \dfrac{LWT}{12}$
Find the number of board feet if $L = 14$ feet, $W = 8$ feet, and $T = 6$ feet.

4. The perimeter of a square is figured by the formula $P = 4s$.
Find the perimeter if $s = 6$.

5. What is the circumference of a circle with a diameter of 8 cm?
$C = \pi d$ $\pi \approx 3.14$

6. Find the area of the trapezoid
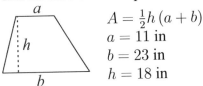
$A = \frac{1}{2}h(a+b)$
$a = 11$ in
$b = 23$ in
$h = 18$ in

7. Find the volume of a sphere with a radius of 6 cm. $\pi \approx 3.14$
$V = \frac{4}{3}\pi r^3$
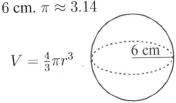

8. Find the area of the following ellipse given by the equation: $A = \pi ab$

$\pi \approx 3.14$
$a = 2$ cm
$b = 4$ cm

9. The formula for changing from degrees Fahrenheit to degrees Celsius is:
$$C = \frac{5(F-32)}{9}$$
If it is 68°F outside, how many degrees Celsius is it?

10. Find the volume. $V = \frac{4}{3}\pi r^3$ $\pi \approx 3.14$

11. Louise has a cone-shaped mold to make candles. The diameter of the base is 10 cm, and it is 13 cm tall. How many cubic centimeters of liquid wax will it hold?
$\pi \approx 3.14$
$V = \frac{1}{3}\pi r^2 h$

Chapter 8   Introduction to Algebra

8.7   Properties of Addition and Multiplication

The associative, commutative, distributive, identity, and inverse properties of addition and multiplication are listed below by example as a quick refresher.

Property	Example
1. Associative Property of Addition	$(a + b) + c = a + (b + c)$
2. Associative Property of Multiplication	$(a \times b) \times c = a \times (b \times c)$
3. Commutative Property of Addition	$a + b = b + a$
4. Commutative Property of Multiplication	$a \times b = b \times a$
5. Distributive Property	$a \times (b + c) = (a \times b) + (a \times c)$
6. Identity Property of Addition	$0 + a = a$
7. Identity Property of Multiplication	$1 \times a = a$
8. Inverse Property of Addition	$a + (-a) = 0$
9. Inverse Property of Multiplication	$a \times \dfrac{1}{a} = \dfrac{a}{a} = 1, a \neq 0$

The reflexive, symmetric, and transitive properties of equality are also listed with examples.

Property	Example
10. Reflexive Property of Equality	$a = a$
11. Symmetric Property of Equality	$a = b$ then $b = a$
12. Transitive Property of Equality (If the first number equals the second, and the second is equal to the third, then the first must be equal to the third.)	If $a = b$ and $b = c$, then $a = c$.

Write the number of the property listed above that describes each of the following statements.

1. $4 + 5 = 5 + 4$ 2

2. $4 + (2 + 8) = (4 + 2) + 8$ 1

3. $10(4 + 7) = (10)(4) + (10)(7)$ 5

4. $(2 \times 3) \times 4 = 2 \times (3 \times 4)$ 2

5. $1 \times 12 = 12$ 7

6. $8\left(\dfrac{1}{8}\right) = 1$ 9

7. $1c = c$ 7

8. If $t = z$ and $z = q$, then $t = q$ 12

9. $42 = 42$ 10

10. $18 + 0 = 18$ 6

11. $9 + (-9) = 0$ 8

12. $p \times q = q \times p$ 3

13. $t + 0 = t$ 6

14. $x(y + z) = xy + xz$ 5

15. $(m)(n \cdot p) = (m \cdot n)(p)$ 2

16. $-y + y = 0$ 8

17. If $a = z$, then $z = a$ 11

18. If $f = g$ and $g = 107$, then $f = 107$ 12

84   Copyright © American Book Company

Chapter 8 Review

Solve the following problems using $x = 2$.

1. $2x - 5 =$
2. $\dfrac{18}{x} =$
3. $x^2 - 4 =$
4. $\dfrac{x^3 + 3}{2} =$
5. $11 - 3x =$
6. $x + 20 =$
7. $-4x + 2 =$
8. $8 - x =$
9. $5x - 2 =$

Solve the following problems. Let $w = -\tfrac{1}{2}, y = 2, z = 3$.

10. $4w - y =$
11. $wyz + 3 =$
12. $z - 4w =$
13. $\dfrac{2z + 1}{wz} =$
14. $\dfrac{8w}{y} + \dfrac{z}{w} =$
15. $30 - 5yz =$
16. $-3y + 2 =$
17. $5w - (yw) =$
18. $8y - 3z =$

Identify the property used in each equation below.

19. $a + 0 = a$
20. $1 \times a = a$
21. $x(y + z) = xy + xz$
22. $a\left(\tfrac{1}{a}\right) = 1$
23. $(m)(n \cdot p) = (m \cdot n)(p)$
24. $-y + y = 0$
25. If $m = n$ and $n = o$, then $m = o$.
26. $a \times b = b \times a$

Write out the algebraic expression given in each word problem.

27. three less than the sum of x and 10

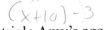

28. triple Amy's age, a

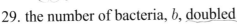

29. the number of bacteria, b, doubled

30. five less than the product of 4 and y

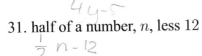

31. half of a number, n, less 12

32. the quotient of a number, x, and 7

Chapter 8 Introduction to Algebra

For questions 33–37, write an equation or expression to match each problem.

33. Calista earns $300 per week for a 40 hour work week plus $12.32 per hour for each hour of overtime after 40 hours. Write an equation that would be used to determine her weekly wages where w is her wages, and v is the number of overtime hours worked.

34. Daniel purchased a 1 year CD, c, from a bank. He bought it at an annual interest rate of 7%. After 1 year, Daniel cashes in the CD. Write an equation that would determine the total amount it is worth.

35. Omar is a salesman. He earns an hourly wage of $12.00 per hour plus he receives a commission of 5% on the sales he makes. Write an equation which would be used to determine his weekly salary, w, where x is the number of hours worked, and y is the amount of sales for the week.

36. Juan sold a boat that he bought 5 years ago. He sold it for 70% less than he originally paid for it. If the original cost is b, write an expression that shows how much he sold the boat for.

37. Toshi is going to get a 5% raise after he works at his job for 1 year. If s represents his starting salary, write an expression that shows how much he will make after his raise.

Answer each of the following questions.

38. Lumber is measured with the following formula:

    Number of board feet $= \dfrac{LWT}{12}$

    $L = $ Length of the board in feet

    $W = $ Width of the board in feet

    $T = $ Thickness of the board in feet

    Find the number of board feet if $L = 12$ feet, $W = 6$ feet, and $T = \frac{1}{4}$ feet.

39. To convert from degrees Celsius to degrees Fahrenheit, use the following formula:

$$F = \frac{9C}{5} + 32$$

    If it is 15°C outside, what is the temperature in degrees Fahrenheit?

86   Copyright © American Book Company

Chapter 8 Test

1. A box has length of 20 inches, the height of 12 inches, and the width of 38 inches. What is the volume of the box using the formula $V = lwh$?

   A. 70 in^3
   B. 240 in^3
   C. 840 in^3
   D. $9,120 \text{ in}^3$
   E. 72 in^3

2. Tom earns \$400 per week before taxes are taken out. His employer takes out a total of 31% for state, federal, and Social Security taxes. Which expression below will help Tom figure his net pay?

   F. $400 - 0.31$
   G. $400 + 0.31$
   H. $400 \times 0.31\,(400)$
   J. $400 + 0.31(400)$
   K. $400 - 0.31\,(400)$

3. Simplify the following expression using $x = 2$ and $y = 5$.

   $3x + 4y - 1$

   A. 22
   B. 13
   C. 25
   D. 10
   E. 24

4. Which property is demonstrated by the expression below?
   $2 + (3 + 8) = (2 + 3) + 8$

   F. commutative property of addition
   G. associative property of addition
   H. distributive property
   J. identity property of addition
   K. inverse property of addition

5. Is the equation $5\,(x - 6) = -2$ equivalent to the equation $5x - 30 = -2$?

   A. Yes, because of the commutative property.
   B. Yes, because of the distributive property.
   C. Yes, because of the associative property.
   D. Yes, because of the inverse property.
   E. Yes, because of the identity property.

6. A plumber charges \$37.00 per hour plus a \$15.00 service charge. If a represents his total charges in dollars, and b represents the number of hours worked, which formula below could the plumber use to calculate his total charges?

   F. $a = 37 + 15b$
   G. $a = 37 + 15 + b$
   H. $a = 37b + 15b$
   J. $a = (37)\,(15) + b$
   K. $a = 37b + 15$

7. Write the expression from the following word problem.

   A number divided by the sum of nine and two.

   A. $\dfrac{x}{9} + 2$
   B. $\dfrac{9 + 2}{x}$
   C. $\dfrac{x}{9 + 2}$
   D. $\dfrac{x + 2}{9}$
   E. $x \div 9 + 2$

Copyright © American Book Company

87

Chapter 8   Introduction to Algebra

8. In 2012, Bell computers announced to its sales force to expect a 2.4% price increase on all computer equipment in the year 2013. A certain sales representative wanted to see how much the increase would be on a computer, c, that sold for $2,600 in 2012. Which expression below will help him find the cost of the computer in the year 2013?

   F. $0.24\,(2,600)$

   G. $2,600 + 0.024\,(2,600)$

   H. $0.24\,(2,600) + 2,600$

   J. $0.024\,(2,600) - 2,600$

   K. $2,600 \div 0.24$

9. $2(3r + 7) =$

   A. $6r + 9$

   B. $5r + 9$

   C. $3r^2 + 14$

   D. $6r + 7$

   E. $6r + 14$

10. In the problem below, $x = 3$, $y = 4$, and $z = 2$. Solve.

   $4z(3y + 2x)$

   F. 18

   G. 144

   H. 142

   J. 124

   K. 112

11. $12(2x - 2) =$

   A. $24x - 2$

   B. $2x - 24$

   C. $24x - 24$

   D. $12x - 12$

   E. $24x - 12$

12. In the problem below, $g = 2$, $h = 5$, and $j = 4$. Solve.

   $2g(5j + 3h)$

   F. 35

   G. 70

   H. 95

   J. 125

   K. 140

13. Which expression demonstrates the commutative property of multiplication?

   A. $(a \times b) \times c = a \times (b \times c)$

   B. $a \times (b + c) = (a \times b) + (a \times c)$

   C. $1 \times a = a$

   D. $a \times b = b \times a$

   E. $a \times \dfrac{1}{a} = \dfrac{a}{a} = 1, a \neq 0$

14. Which expression demonstrates the transitive property of equality?

   F. $a = a$

   G. $a = b$ then $b = a$

   H. If $a = b$ and $b = c$, then $a = c$.

   J. If $a = b$ and $b \neq c$, then $a = c$

   K. $a = b$ then $b \neq a$

15. $5(2x - 6) =$

   A. $10x - 30$

   B. $10x - 6$

   C. $2x - 30$

   D. $5 + 2x - 6$

   E. $25x - 65$

Chapter 9
Equations and Inequalities

9.1   Two-Step Algebra Problems

In the following two-step algebra problems, **additions** and **subtractions** are performed first and then **multiplication** and **division**.

Example 1: $-4x + 7 = 31$

Step 1:   Subtract 7 from both sides.

$$
\begin{array}{rl}
-4x + 7 & = 31 \\
-7 & -7 \\
\hline
-4x & = 24
\end{array}
$$

Step 2:   Divide both sides by -4.

$$\frac{-4x}{-4} = \frac{24}{-4} \qquad \text{so } x = -6$$

Example 2: $-8 - y = 12$

Step 1:   Add 8 to both sides.

$$
\begin{array}{rl}
-8 - y & = 12 \\
+8 & +8 \\
\hline
-y & = 20
\end{array}
$$

Step 2:   To finish solving a problem with a negative sign in front of the variable, multiply both sides by -1. The variable needs to be positive in the answer.

$$(-1)(-y) = (-1)(20) \text{ so } y = -20$$

Copyright © American Book Company

89

Chapter 9   Equations and Inequalities

Solve the two-step algebra problems below.

1. $6x - 4 = -34$

2. $5y - 3 = 32$

3. $8 - t = 1$

4. $10p - 6 = -36$

5. $11 - 9m = -70$

6. $4x - 12 = 24$

7. $3x - 17 = -41$

8. $9d - 5 = 49$

9. $10h + 8 = 78$

10. $-6b - 8 = 10$

11. $-g - 24 = -17$

12. $-7k - 12 = 30$

13. $9 - 5r = 64$

14. $6y - 14 = 34$

15. $12f + 15 = 51$

16. $21t + 17 = 80$

17. $20y + 9 = 149$

18. $15p - 27 = 33$

19. $22h + 9 = 97$

20. $-5 + 36w = 175$

9.2   Two-Step Algebra Problems with Fractions

An algebra problem may contain a fraction. Study the following example to understand how to solve algebra problems that contain a fraction.

Example 3: $\dfrac{x}{2} + 4 = 3$

Step 1:

$$\dfrac{x}{2} + 4 = 3$$
$$\underline{-4 \qquad -4}$$
$$\dfrac{x}{2} \qquad = -1$$

Subtract 4 from both sides.

Step 2: $\dfrac{x}{2} = -1$ Multiply both sides by 2 to eliminate the fraction.

$$\dfrac{x}{\cancel{2}} \times \cancel{2} = -1 \times 2, \ x = -2$$

Simplify the following algebra problems.

1. $4 + \dfrac{y}{3} = 7$

2. $\dfrac{a}{2} + 5 = 12$

3. $\dfrac{w}{5} - 3 = 6$

4. $\dfrac{x}{9} - 9 = -5$

5. $\dfrac{b}{6} + 2 = -4$

6. $7 + \dfrac{z}{2} = -13$

7. $\dfrac{x}{2} - 7 = 3$

8. $\dfrac{c}{5} + 6 = -2$

9. $3 + \dfrac{x}{11} = 7$

10. $16 + \dfrac{m}{6} = 14$

11. $\dfrac{p}{3} + 5 = -2$

12. $\dfrac{t}{8} + 9 = 3$

13. $\dfrac{v}{7} - 8 = -1$

14. $5 + \dfrac{h}{10} = 8$

15. $\dfrac{k}{7} - 9 = 1$

16. $\dfrac{y}{4} + 13 = 8$

17. $15 + \dfrac{z}{14} = 13$

18. $\dfrac{b}{6} - 9 = -14$

19. $\dfrac{d}{3} + 7 = 12$

20. $10 + \dfrac{b}{6} = 4$

21. $2 + \dfrac{p}{4} = -6$

22. $\dfrac{t}{7} - 9 = -5$

23. $\dfrac{a}{10} - 1 = 3$

24. $\dfrac{a}{8} + 16 = 9$

90   Copyright © American Book Company

9.3   More Two-Step Algebra Problems with Fractions

9.3   More Two-Step Algebra Problems with Fractions

Study the following example to understand how to solve algebra problems that contain a different type of fraction.

Example 4: $\dfrac{x+2}{4} = 3$ In this example, "$x+2$" is divided by 4, and not just the x or the 2.

Step 1: $\dfrac{x+2}{4} \times 4 = 3 \times 4$ First, multiply both sides by 4 to eliminate the fraction.

Step 2:
$$
\begin{aligned}
x + 2 &= 12 \\
-2 \quad\; &\;\; -2 \\
\hline
x &= 10
\end{aligned}
$$
Next, subtract 2 from both sides.

Solve the following problems.

1. $\dfrac{x+1}{5} = 4$

2. $\dfrac{z-9}{2} = 7$

3. $\dfrac{b-4}{4} = -5$

4. $\dfrac{y-9}{3} = 7$

5. $\dfrac{d-10}{-2} = 12$

6. $\dfrac{w-10}{-8} = -4$

7. $\dfrac{x-1}{-2} = -5$

8. $\dfrac{c+40}{-5} = -7$

9. $\dfrac{13+h}{2} = 12$

10. $\dfrac{k-10}{3} = 9$

11. $\dfrac{a+11}{-4} = 4$

12. $\dfrac{x-20}{7} = 6$

13. $\dfrac{t+2}{6} = -5$

14. $\dfrac{b+1}{-7} = 2$

15. $\dfrac{f-9}{3} = 8$

16. $\dfrac{4+w}{6} = -6$

17. $\dfrac{3+t}{3} = 10$

18. $\dfrac{x+5}{5} = -3$

19. $\dfrac{g+3}{2} = 11$

20. $\dfrac{k+1}{-6} = 5$

21. $\dfrac{y-14}{2} = -8$

22. $\dfrac{z-4}{-2} = 13$

23. $\dfrac{w+2}{15} = -1$

24. $\dfrac{3+h}{3} = 6$

Copyright © American Book Company

Chapter 9   Equations and Inequalities

9.4   Combining Like Terms

In algebra problems, **terms** are often separated by $+$ and $-$ signs. The expression $5x - 4 - 3x + 7$ has 4 terms: $5x$, 4, $3x$, and 7. Terms having the same variable can be combined (added or subtracted) to simplify the expression. $5x - 4 - 3x + 7$ simplifies to $2x + 3$.

$$5x - 3x \quad - 4 + 7 \ = 2x + 3$$

Simplify the following expressions.

1. $7x + 12x$

2. $8y - 5y + 8$

3. $4 - 2x + 9$

4. $11a - 16 - a$

5. $9w + 3w + 3$

6. $-5x + x + 2x$

7. $w - 15 + 9w$

8. $21 - 10t + 9 - 2t$

9. $-3 + x - 4x + 9$

10. $7b + 12 + 4b$

11. $4h - h + 2 - 5$

12. $-6k + 10 - 4k$

13. $2a + 12a - 5 + a$

14. $5 + 9c - 10$

15. $-d + 1 + 2d - 4$

16. $-8 + 4h + 1 - h$

17. $12x - 4x + 7$

18. $10 + 3z + z - 5$

19. $14 + 3y - y - 2$

20. $11p - 4 + p$

21. $11m + 2 - m + 1$

9.5   Solving Equations with Like Terms

When an equation has two or more like terms on the same side of the equation, combine like terms as the **first** step in solving the equation.

Example 5: $7x + 2x - 7 = 21 + 8$

Step 1:   Combine like terms on both sides of the equation.

$$
\begin{aligned}
7x + 2x - 7 &= 21 + 8 \\
9x - 7 &= 29 \\
+7 \quad &\quad +7 \\
9x \div 9 &= 36 \div 9 \\
x &= 4
\end{aligned}
$$

Step 2:   Solve the two-step algebra problem as explained previously. In this case, add 7 to both sides, then divide both sides by 9.

Solve the equations below combining like terms first.

1. $3w - 2w + 4 = 6$

2. $7x + 3 + x = 16 + 3$

3. $5 - 6y + 9y = -15 + 5$

4. $-14 + 7a + 2a = -5$

5. $-2t + 4t - 7 = 9$

6. $9d + d - 3d = 14$

7. $-6c - 4 - 5c = 10 + 8$

8. $15m - 9 - 6m = 9$

9. $-4 - 3x - x = -16$

10. $9 - 12p + 5p = 14 + 2$

11. $10y + 4 - 7y = -17$

12. $-8a - 15 - 4a = 9$

92   Copyright © American Book Company

9.5   Solving Equations with Like Terms

If the equation has like terms on both sides of the equation, you must get all of the terms with a **variable** on one side of the equation and all of the **integers** on the other side of the equation.

Example 6: $3x + 2 = 6x - 1$

Step 1:	Subtract $6x$ from both sides to move all the **variables** to one side.	
Step 2:	Subtract 2 from both sides to move all the **integers** to the other side.	
Step 3:	Divide by -3 to solve for x.	

$$\begin{array}{rl} 3x + 2 &= 6x - 1 \\ -6x & -6x \\ \hline -3x + 2 &= -1 \\ -2 & -2 \\ \hline \dfrac{-3x}{-3} &= \dfrac{-3}{-3} \\ x &= 1 \end{array}$$

Solve the following problems.

1. $3a + 1 = a + 9$

2. $2d - 12 = d + 3$

3. $5x + 6 = 14 - 3x$

4. $15 - 4y = 2y - 3$

5. $9w - 7 = 12w - 13$

6. $10b + 19 = 4b - 5$

7. $-7m + 9 = 29 - 2m$

8. $5x - 26 = 13x - 2$

9. $19 - p = 3p - 9$

10. $-7p - 14 = -2p + 11$

11. $16y + 12 = 9y + 33$

12. $13 - 11w = 3 - w$

13. $-17b + 23 = -4 - 8b$

14. $k + 5 = 20 - 2k$

15. $12 + m = 4m + 21$

16. $7p - 30 = p + 6$

17. $19 - 13z = 9 - 12z$

18. $8y - 2 = 4y + 22$

19. $5 + 16w = 6w - 45$

20. $-27 - 7x = 2x + 18$

21. $-12x + 14 = 8x - 46$

22. $27 - 11h = 5 - 9h$

23. $5t + 36 = -6 - 2t$

24. $17y + 42 = 10y + 7$

25. $22x - 24 = 14x - 8$

26. $p - 1 = 4p + 17$

27. $4d + 14 = 3d - 1$

28. $7w - 5 = 8w + 12$

29. $-3y - 2 = 9y + 22$

30. $17 - 9m = m - 23$

Copyright © American Book Company   93

Chapter 9   Equations and Inequalities

9.6   Solving for a Variable

Sometimes an equation has two variables, and you may be asked to solve for one of the variables.

Example 7:   If $5x + y = 19$, then $y =$

Solution: The goal is to have only y on one side of the equation and the rest of the terms on the other side of the equation. Follow the order of operations to solve the problem.

$5x + y - 5x = 19 - 5x$ Subtract $5x$ from both sides of the equation.
$y = 19 - 5x$

Example 8:   If $7m + n = 30$, then $m =$

Solution: The goal is to have only m on one side of the equation and the rest of the terms on the other side of the equation. Follow the order of operations to solve the problem.

$7m + n = 30$ Subtract n from both sides of the equation.

$7m + n - n = 30 - n$

$\dfrac{7m}{7} = \dfrac{30 - n}{7}$ Divide both sides of the equation by 7.

$m = \dfrac{30 - n}{7}$

Solve each of the equations below for the variable indicated. Be sure to follow the order of operations.

1. If $4a + b = 12$, then $a =$

2. If $6c - d = 17$, then $d =$

3. If $3m - n = 11$, then $m =$

4. If $7r + 5s = 35$, then $r =$

5. If $8m - 9n - 2m = 6n$, then $m =$

6. If $-10y - 3x - x = -12y$, then $x =$

7. If $-4t + 4t - 2s = 8$, then $s =$

8. If $5x - 7y + 9y = -9x + 3$, then $y =$

9. If $-10b + 7a + 3a = -4b$, then $a =$

10. If $7x - 8y + x = 5y + 3$, then $x =$

94   Copyright © American Book Company

9.7 Removing Parentheses

The distributive property is used to remove parentheses.

Example 9: $2(a+6)$

You multiply 2 by each term inside the parentheses. $2 \times a = 2a$ and $2 \times 6 = 12$. The 12 is a positive number so use a plus sign between the terms in the answer.
$2(a+6) = 2a + 12$

Example 10: $4(-5c+2)$

The first term inside the parentheses could be negative. Multiply in exactly the same way as the example above. $4 \times (-5c) = -20c$ and $4 \times 2 = 8$
$4(-5c+2) = -20c + 8$

Use the distributive property to remove the parentheses in the problems below.

1. $7(n+6)$
2. $8(2g-5)$
3. $11(5z-2)$
4. $6(-y-4)$
5. $3(-3k+5)$

6. $4(d-8)$
7. $2(-4x+6)$
8. $7(4+6p)$
9. $5(-4w-8)$
10. $6(11x+2)$

11. $10(9-y)$
12. $9(c-9)$
13. $12(-3t+1)$
14. $3(4y+9)$
15. $8(b+3)$

The number in front of the parentheses can also be negative. Remove these parentheses the same way.

Example 11: $-2(b-4)$

First, multiply $-2 \times b = -2b$
Second, multiply $-2 \times -4 = 8$
Copy the two products. The second product is a positive number so put a plus sign between the terms in the answer.
$-2(b-4) = -2b + 8$

Use the distributive property to remove the parentheses in the following problems.

16. $-7(x+2)$
17. $-5(4-y)$
18. $-4(2b-2)$
19. $-2(8c+6)$
20. $-5(-w-8)$

21. $-3(4x-2)$
22. $-2(-z+2)$
23. $-4(7p+7)$
24. $-9(t-6)$
25. $-10(2w+4)$

26. $-3(9-7p)$
27. $-9(-k-3)$
28. $-1(7b-9)$
29. $-6(-5t-2)$
30. $-7(-v+4)$

Copyright © American Book Company

Chapter 9   Equations and Inequalities

9.8   Multi-Step Algebra Problems

You can now use what you know about removing parentheses, combining like terms, and solving simple algebra problems to solve problems that involve three or more steps. Study the examples below to see how to solve multi-step problems.

Example 12: $3(x+6) = 5x - 2$

Step 1:	Use the distributive property to remove parentheses.	$3x + 18 = 5x - 2$
Step 2:	Subtract $5x$ from each side to move the terms with variables to the left side of the equation.	$\underline{ -5x -5x}$ $-2x + 18 = -2$
Step 3:	Subtract 18 from each side to move the integers to the right side of the equation.	$\underline{ -18 -18}$ $\dfrac{-2x}{-2} = \dfrac{-20}{-2}$
Step 4:	Divide both sides by -2 to solve for x.	$x = 10$

Example 13: $\dfrac{3(x-3)}{2} = 9$

Step 1:	Use the distributive property to remove parentheses.	$\dfrac{3x - 9}{2} = 9$
Step 2:	Multiply both sides by 2 to eliminate the fraction.	$\dfrac{2(3x - 9)}{2} = 2(9)$
Step 3:	Add 9 to both sides, and combine like terms.	$3x - 9 = 18$ $\underline{ +9 +9}$
Step 4:	Divide both sides by 3 to solve for x.	$\dfrac{3x}{3} = \dfrac{27}{3}$ $x = 9$

Solve the following multi-step algebra problems.

1. $2(y-3) = 4y + 6$

2. $\dfrac{2(a+4)}{2} = 12$

3. $\dfrac{10(x-2)}{5} = 14$

4. $\dfrac{12y - 18}{6} = 4y + 3$

5. $2x + 3x = 30 - x$

6. $\dfrac{2a+1}{3} = a + 5$

7. $5(b-4) = 10b + 5$

8. $-8(y+4) = 10y + 4$

96   Copyright © American Book Company

9.8 Multi-Step Algebra Problems

9. $\dfrac{x+4}{-3} = 6 - x$

10. $\dfrac{4(n+3)}{5} = n - 3$

11. $3(2x-5) = 8x - 9$

12. $7 - 10a = 9 - 9a$

13. $7 - 5x = 10 - (6x + 7)$

14. $4(x-3) - x = x - 6$

15. $4a + 4 = 3a - 4$

16. $-3(x-4) + 5 = -2x - 2$

17. $5b - 11 = 13 - b$

18. $\dfrac{-4x+3}{2x} = \dfrac{7}{2x}$

19. $-(x+1) = -2(5-x)$

20. $4(2c+3) - 7 = 13$

21. $6 - 3a = 9 - 2(2a+5)$

22. $-5x + 9 = -3x + 11$

23. $3y + 2 - 2y - 5 = 4y + 3$

24. $3y - 10 = 4 - 4y$

25. $-(a+3) = -2(2a+1) - 7$

26. $5m - 2(m+1) = m - 10$

27. $\dfrac{1}{2}(b-2) = 5$

28. $-3(b-4) = -2b$

29. $4x + 12 = -2(x+3)$

30. $\dfrac{7x+4}{3} = 2x - 1$

31. $9x - 5 = 8x - 7$

32. $7x - 5 = 4x + 10$

33. $\dfrac{4x+8}{2} = 6$

34. $2(c+4) + 8 = 10$

35. $y - (y+3) = y + 6$

36. $4 + x - 2(x-6) = 8$

Copyright © American Book Company

Chapter 9   Equations and Inequalities

9.9   Graphing Inequalities on a Number Line

An inequality is a sentence that contains a $<, >, \le$, or $\ge$ sign. Look at the following graphs of inequalities on a number line.

NUMBER LINE

$x < 3$ is read "x is less than 3."

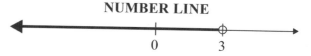

There is no line under the $<$ sign, so the graph uses an **open** endpoint to show x is less than 3 but does not include 3. All the numbers less than 3 are shaded.

$x \le 5$ is read "x is less than or equal to 5."

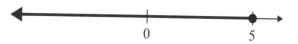

If you see a line under $<$ or $>$ ($\le$ or $\ge$), the endpoint is filled in. The graph uses a **closed** circle because the number 5 is included in the graph.

$x > -2$ is read "x is greater than -2."

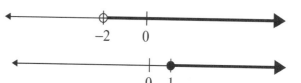

$x \ge 1$ is read "x is greater than or equal to 1."

There can be more than one inequality sign. This is called a **compound inequality**. For example:

$-2 \le x < 4$ is read "-2 is less than or equal to x and x is less than 4."

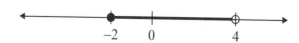

Graph the solution sets of the following inequalities.

1. $x > 8$
2. $x \le 5$
3. $-5 < x < 1$
4. $x > 7$
5. $1 \le x < 4$
6. $x \ge 10$

Give the inequality represented by each of the following number lines.

7.  _____

8. _____

9. _____

10.  _____

11. _____

12. _____

9.10 Multi-Step Inequalities

Remember that adding and subtracting with inequalities follow the same rules as equations. When you multiply or divide both sides of an inequality by the same positive number, the rules are the same as for equations. However, when you multiply or divide both sides of an inequality by a **negative** number, you must **reverse** the inequality symbol.

Example 14:
$$-x > 4$$
$$(-1)(-x) < (-1)(4)$$
$$x < -4$$

Example 15:
$$-4x < 2$$
$$\frac{-4x}{-4} > \frac{2}{-4}$$
$$x > -\frac{1}{2}$$

Reverse the symbol when you multiply or divide by a negative number.

When solving multi-step inequalities, first add and subtract to isolate the term with the variable. Then multiply and divide.

Example 16: $2x - 8 > 4x + 1$

Step 1: Add 8 to both sides.

$$2x - 8 + 8 > 4x + 1 + 8$$
$$2x > 4x + 9$$

Step 2: Subtract $4x$ from both sides.

$$2x - 4x > 4x + 9 - 4x$$
$$-2x > 9$$

Step 3: Divide by -2. Remember to change the direction of the inequality sign.

$$\frac{-2x}{-2} < \frac{9}{-2}$$
$$x < -\frac{9}{2}$$

Copyright © American Book Company

Chapter 9   Equations and Inequalities

Solve each of the following inequalities.

1. $8 - 3x \leq 7x - 2$

2. $3(2x - 5) \geq 8x - 5$

3. $\dfrac{1}{3}b - 2 > 5$

4. $7 + 3y > 2y - 5$

5. $3a + 5 < 2a - 6$

6. $3(a - 2) > -5a - 2(3 - a)$

7. $2x - 7 \geq 4(x - 3) + 3x$

8. $6x - 2 \leq 5x + 5$

9. $-\dfrac{x}{4} > 12$

10. $-\dfrac{2x}{3} \leq 6$

11. $3b + 5 < 2b - 8$

12. $4x - 5 \leq 7x + 13$

13. $4x + 5 \leq -2$

14. $2y - 5 > 7$

15. $4 + 2(3 - 2y) \leq 6y - 20$

16. $-4c + 6 \leq 8$

17. $-\dfrac{1}{2}x + 2 > 9$

18. $\dfrac{1}{4}y - 3 \leq 1$

19. $-3x + 4 > 5$

20. $\dfrac{y}{2} - 2 \geq 10$

21. $7 + 4c < -2$

22. $2 - \dfrac{a}{2} > 1$

23. $10 + 4b \leq -2$

24. $-\dfrac{1}{2}x + 3 > 4$

9.11 Number Patterns

In each of the examples below, there is a sequence of numbers that follows a pattern. Think of the sequence of numbers like the output for a function. You must find the pattern (or function) that holds true for each number in the sequence. Once you determine the pattern, you can find the next number in the sequence or any number in the sequence.

	Sequence	Pattern	Next Number	20th number in the sequence
Example 17:	$3, 4, 5, 6, 7$	$n + 2$	8	22

In number patterns, the sequence is the output. The input can be the set of whole numbers starting with 1. But, you must determine the "rule" or pattern. Look at the table below.

input	sequence
1 $\longrightarrow$	3
2 $\longrightarrow$	4
3 $\longrightarrow$	5
4 $\longrightarrow$	6
5 $\longrightarrow$	7

What pattern or "rule" can you come up with that gives you the first number in the sequence, 3, when you input 1? $n + 2$ will work because when $n = 1$, the first number in the sequence $= 3$. Does this pattern hold true for the rest of the numbers in the sequence? Yes, it does. When $n = 2$, the second number in the sequence $= 4$. When $n = 3$, the third number in the sequence $= 5$, and so on. Therefore, $n + 2$ is the pattern. Even without knowing the algebraic form of the pattern, you could determine that 8 is the next pattern in the sequence. To find the 20th number in the pattern, use $n = 20$ to get 22.

	Sequence	Pattern	Next Number	20th number in the sequence
Example 18:	$1, 4, 9, 16, 25$	n^2	36	400
Example 19:	$-2, -4, -6, -8, -10$	$-2n$	-12	-40

Find the pattern, the next number, and the 20th number in each of the sequences below.

	Sequence	Pattern	Next Number	20th number in the sequence
1.	$-2, -1, 0, 1, 2$			
2.	$5, 6, 7, 8, 9$			
3.	$3, 7, 11, 15, 19$			
4.	$-3, -6, -9, -12, -15$			
5.	$3, 5, 7, 9, 11$			
6.	$2, 4, 8, 16, 32$			
7.	$1, 8, 27, 64, 125$			
8.	$0, -1, -2, -3, -4$			
9.	$2, 5, 10, 17, 26$			
10.	$4, 6, 8, 10, 12$			

Copyright © American Book Company

Chapter 9   Equations and Inequalities

Chapter 9 Review

Solve each of the following equations.

1. $4a - 8 = 28$

3. $-7 + 23w = 108$

5. $c - 13 = 5$

2. $5 + \dfrac{x}{8} = -4$

4. $\dfrac{y - 8}{6} = 7$

6. $\dfrac{b + 9}{12} = -3$

Simplify the following expressions by combining like terms.

7. $-4a + 8 + 3a - 9$

8. $14 + 2z - 8 - 5z$

9. $-7 - 7x - 2 - 9x$

Solve.

10. $19 - 8d = d - 17$

12. $6 + 16x = -2x - 12$

14. $4x - 16 = 7x + 2$

11. $7w - 8w = -4w - 30$

13. $6(b - 4) = 8b - 18$

15. $9w - 2 = -w - 22$

Use the distributive property to remove parentheses.

16. $3(-4x + 7)$

18. $6(8 - 9b)$

20. $-2(5c - 3)$

17. $11(2y + 5)$

19. $-8(-2 + 3a)$

21. $-5(7y - 1)$

Solve for the indicated variable.

22. If $3x - y = 15$, then $y =$

23. If $7a + 2b = 1$, then $b =$

Solve each of the following equations and inequalities.

24. $\dfrac{-11c - 35}{4} = 4c - 2$

28. $\dfrac{5(n + 4)}{3} = n - 8$

25. $5 + x - 3(x + 4) = -17$

29. $-y > 14$

26. $4(2x + 3) \geq 2x$

30. $2(3x - 1) \geq 3x - 7$

27. $7 - 3x \leq 6x - 2$

31. $3(x + 2) < 7x - 10$

Find the pattern for the following number sequences, and then find the nth number requested.

32. 0, 1, 2, 3, 4   pattern_____

35. 1, 3, 5, 7, 9   25th number_____

33. 0, 1, 2, 3, 4   20th number_____

36. 3, 6, 9, 12, 15   pattern_____

34. 1, 3, 5, 7, 9   pattern_____

37. 3, 6, 9, 12, 15   30th number_____

102   Copyright © American Book Company

Chapter 9 Test

1. Find the value of n. $19n - 57 = 76$

   A. 1
   B. 3
   C. 5
   D. 7
   E. 6

2. Solve for x. $14x + 84 = 154$

   F. 4
   G. 5
   H. 11
   J. 17
   K. 15

3. Which of the following is equivalent to $4 - 5x = 3(x - 4)$?

   A. $4 - 5x = 3x - 4$
   B. $4 - 5x = 3x + 1$
   C. $4 - 5x = 3x - 1$
   D. $4 - 5x = 3x - 7$
   E. $4 - 5x = 3x - 12$

4. Which of the following is equivalent to $3(x - 2) + 1 - 2x = -4$?

   F. $x - 6 = -4$
   G. $-6x + 1 = -4$
   H. $5x - 7 = -4$
   J. $x - 5 = -4$
   K. $x + 5 = -4$

5. Solve: $3(x - 2) - 1 = 6(x + 5)$

   A. -4
   B. $-\dfrac{37}{3}$
   C. 4
   D. $\dfrac{23}{3}$
   E. $-7\dfrac{2}{3}$

6. $5(2x + 11) - 3 \times 5 = ?$

   F. $7x + 40$
   G. $7x + 20$
   H. $10x + 40$
   J. $10x + 260$
   K. $7x + 45$

7. Solve: $4b - 8 = 56$

   A. $b = 12$
   B. $b = 16$
   C. $b = -12$
   D. $b = -16$
   E. $b = -8$

8. Which of the following is equivalent to $3(2x - 5) - 4(x - 3) = 7$?

   F. $x + 27 = 7$
   G. $2x - 3 = 7$
   H. $10x - 27 = 7$
   J. $x - 27 = 7$
   K. $2x + 3 = 7$

9. What is the next number in this sequence?

   $0.03, 0.12, 0.48, 1.92,$ _____

   A. 1.95
   B. 3.36
   C. 5.08
   D. 7.48
   E. 7.68

10. Which inequality correctly depicts the following graph?

    F. $x < 5$
    G. $x \leq 5$
    H. $x \geq 5$
    J. $x > 5$
    K. $x = 5$

Chapter 10
Algebra Word Problems

10.1   Algebra Word Problems

An equation states that two mathematical expressions are equal. In working with word problems, the words that mean equal are **equals, is, was, is equal to, amounts to,** and other expressions with the same meaning. To translate a word problem into an algebraic equation, use a variable to represent the unknown or unknowns you are looking for.

In the following example, let n be the number you are looking for.

Example 1:   Four more than twice a number is two less than three times the number.

$$
\begin{array}{rcl}
\textbf{Step 1:} \quad \textbf{Translation:} & 4 + 2n & = & 3n - 2 \\
\textbf{Step 2:} \quad \textbf{Now Solve:} & -2n & & -2n \\
\hline
& 4 & = & n - 2 \\
& +2 & & +2 \\
\hline
& 6 & = & n
\end{array}
$$

The number is 6. Substitute the number back into the original equation to check.

Translate the following word problems into equations and solve.

1. Four less than twice a number is ten. Find the number.

2. Three more than three times a number is one less than two times the number. What is the number?

3. The sum of seven times a number and the number is 24. What is the number?

4. Negative 18 is the sum of five and a number. Find the number.

5. Negative 14 is equal to ten minus the product of six and a number. What is the number?

6. Two less than twice a number equals the number plus 12. What is the number?

7. The difference between three times a number and 31 is two. What is the number?

8. Sixteen is fourteen less than the product of a number and five. What is the number?

9. Eight more than twice a number is four times the difference between five and the number. What is the number?

10. Three less than twice a number is three times the sum of one and the number. What is the number?

10.2   Real-World Linear Equations

Linear equations are very useful mathematical tools. They allow us to show relationships between two variables.

Example 2:   A local cell phone company uses the equation $y = \frac{5}{2}x + 10$ to determine the charges for usage where $y =$ the cost and $x =$ the minutes used. How much will Jessica's bill be if she talked for 40 minutes?

Step 1:   Substitute the known value in for x.
$y = \frac{5}{2}(40) + 10$

Step 2:   Simplify.
$y = 100 + 10 = 110$
Jessica's bill will be $110.

Example 3:   Vincent bought a luxury car for $165,000$ and its value has depreciated linearly. After 5 years the value was $137,000$. What is the amount of yearly depreciation?

Step 1:   First, find how much the car's value depreciated in 5 years.
$165,000 - \$137,000 = \$28,000$

Step 2:   Next, find the average yearly depreciation by dividing $28,000$ by the amount of years, 5.
$28,000 \div 5 = \$5,600$
The value of Vincent's car depreciated an average of $5,600$ each year.

Example 4:   In 1990, the average cost of a new house was $123,000$. By the year 2000, the average cost of a new house was $134,150$. Based on a linear model, what is the predicted average cost for 2008?

Step 1:   First, we need to find the difference between the average cost of a new house in the year 1990 and the average cost of a new house in the year 2000.
$134,150 - \$123,000 = \$11,150$

Step 2:   Next, we need to find how much the average cost of a new house went up each year. Since it had been 10 years, divide the difference between the value in 2000 and 1990 by 10.
$11,150 \div 10 = \$1,115$

Step 3:   Multiply the amount the average cost of a new house went up each year by the number of years between 2000 and 2008.
$1,115 \times 8 = \$8,920$

Step 4:   Lastly, add the average cost of a new house in the year 2000 with the amount found in step 3.
$134,150 + \$8,920 = 143,070$
$143,070$ is the predicted average cost of a new house for 2008.

Copyright © American Book Company

Chapter 10   Algebra Word Problems

Solve the following problems.

1. Acacia bought an MP3 player at Everywhere Electronics for $350 and its valued depreciated linearly. Three years later, she saw the same MP3 player at Everywhere Electronics for $125. What is the amount of yearly depreciation of Acacia's MP3 player?

2. Dustin bought a boat 10 years ago for $10,000. Its value depreciated linearly and now it is worth $2,500. What is the amount of yearly depreciation of Dustin's boat?

3. A small plane costs $500,000 new. Twenty years later it is valued at $150,000. Assuming a linear depreciation, what was the value of the plane when it was 14 years old?

4. In 1980, the price of a scientific calculator was $155. In 2005, the price was $15 dollars. Assuming the change in price was linear, what was the price of a scientific calculator in 1997?

5. In 1997, Justin bought a house for $120,000. In 2004, his house was worth $176,000. Based on a linear model, how much was Justin's house worth in 2001?

6. The attendance on the first day of the Sunny Day Festival was 325 people. The attendance on the third day was 383 people. Assuming the attendance increase is linear, how many people will attend the Sunny Day Festival on the seventh day?

7. Two years ago Juanita bought 2 shirts for $15 and last year she bought 4 shirts for $45. Assuming the price increase is linear, how much will 8 shirts cost Juanita this year?

8. In 1985, the average price of a new car was $9,000. In 2000, the average price was $24,750. Based on a linear model, what is the predicted average price for 2009?

Use the following information for questions 9–10.

Abbey is looking for a new cell phone provider. In her search, she has found 3 local companies: Gift of Gab, On the Go, and Connect. To determine their monthly charges, the 3 companies use the following equations.

Gift of Gab: $y = \frac{3}{4}x + 20$

On the Go: $y = \frac{1}{2}x + 60$

Connect: $y = 6x - 100$

9. Which is the cheapest provider if Abbey uses 200 minutes per month?

10. What if she used 100 minutes?

Copyright © American Book Company

10.3 Word Problems with Formulas

The perimeter of a geometric figure is the distance around the outside of the figure.

Example 5: The perimeter of a rectangle is 44 feet. The length of the rectangle is 6 feet more than the width. What is the measure of the width?

Step 1: Let the variable be the length of the unknown side.
width $= w$ length $= 6 + w$

Step 2: Use the equation for the perimeter of a rectangle as follows:
$2l + 2w =$ perimeter of a rectangle
$2(w + 6) + 2w = 44$

Step 3: Solve for w.

Solution: width $= 8$ feet

Example 6: The perimeter of a triangle is 26 feet. The second side is twice as long as the first. The third side is 1 foot longer than the second side. What is the length of the 3 sides?

Step 1: Let $x =$ first side $2x =$ second side $2x + 1 =$ third side

Step 2: Use the equation for perimeter of a triangle as follows:
sum of the length of the sides = perimeter of a triangle.
$x + 2x + 2x + 1 = 26$

Step 3: Solve for x. $5x + 1 = 26$ so $x = 5$

Solution: first side $x = 5$ second side $2x = 10$ third side $2x + 1 = 11$

Solve the following word problems.

1. The length of a rectangle is 4 times longer than the width. The perimeter is 30. What is the width?

2. The length of a rectangle is 3 more than twice the width. The perimeter is 36. What is the length?

3. The perimeter of a triangle is 18 feet. The second side is two feet longer than the first. The third side is two feet longer then the second. What are the lengths of the sides?

4. In an isosceles triangle, two sides are equal. The third side is two less than twice the length of the sum of the two sides. The perimeter is 40. What are the lengths of the three sides?

5. The sum of the measures of the angles of a triangle is $180°$. The second angle is three times the measure of the first angle. The third angle is four times the measure of the second angle. Find the measure of each angle.

6. The sum of the measures of the angles of a triangle is $180°$. The second angle of a triangle is twice the measure of the first angle. The third angle is 20 more than 5 times the first. What are the measures of the three angles?

Copyright © American Book Company

Chapter 10   Algebra Word Problems

10.4   Age Problems

Example 7:   Tara is twice as old as Gwen. Their sister, Amy, is 5 years older than Gwen. If the sum of their ages is 29 years, find each of their ages.

Step 1:   We want to find each of their ages so there are three unknowns. Tara is twice as old as Gwen, and Amy is older than Gwen, so Gwen is the youngest. Let x be Gwen's age. From the problem we can see that:

$$\left.\begin{array}{rcl} \text{Gwen} & = & x \\ \text{Tara} & = & 2x \\ \text{Amy} & = & x+5 \end{array}\right\} \text{The sum of their ages is 29.}$$

Step 2:   Set up the equation, and solve for x.

$$\begin{array}{rcl} x + 2x + x + 5 & = & 29 \\ 4x + 5 & = & 29 \\ 4x & = & 29 - 5 \\ x & = & \dfrac{24}{4} \\ x & = & 6 \end{array}$$

Solution:
$$\begin{array}{rcl} \text{Gwen's age } (x) & = & 6 \\ \text{Tara's age } (2x) & = & 12 \\ \text{Amy's age } (x+5) & = & 11 \end{array}$$

Solve the following age problems.

1. Carol is 15 years older than her cousin Amanda. Cousin Bill is 4 times as old as Amanda. The sum of their ages is 99. Find each of their ages.

2. Derrick is 5 less than twice as old as Brandon. The sum of their ages is 31. How old are Derrick and Brandon?

3. Beth's mom is 5 times older than Beth. Beth's dad is 8 years older than Beth's mom. The sum of their ages is 74. How old are each of them?

4. Delores is 4 years more than three times as old as her son, Raul. If the difference between their ages is 34, how old are Delores and Raul?

5. Eileen is 9 years older than Karen. John is three times as old as Karen. The sum of their ages is 64. How old are Eileen, Karen, and John?

6. Taylor is 20 years younger than Jim. Andrew is twice as old as Taylor. The sum of their ages is 32. How old are Taylor, Jim, and Andrew?

108   Copyright © American Book Company

10.4   Age Problems

The following problems work in the same way as the age problems. There are two or three items of different weight, distance, number, or size. You are given the total and asked to find the amount of each item.

7. Three boxes have a total weight of 720 pounds. Box A weighs twice as much as Box B. Box C weighs 30 pounds more than Box A. How much do each of the boxes weigh?

8. There are 170 students registered for American History classes. There are twice as many students registered in second period as first period. There are 10 less than three times as many students registered in third period as in first period. How many students are in each period?

9. Mei earns $4 less than three times as much as Olivia. Shane earns twice as much as Mei. Together they earn $468 per week. How much does each person earn per week?

10. Ellie, the elephant, eats 4 times as much as Popcorn, the pony. Zac, the zebra, eats twice as much as Popcorn. Altogether, they eat 238 kilograms of feed per week. How much feed does each of them require each week?

11. The school cafeteria served three kinds of lunches today to 117 students. The students chose the cheeseburgers three times more often than the grilled cheese sandwiches. There were twice as many grilled cheese sandwiches sold as fish sandwiches. How many of each lunch were served?

12. Three friends drove southeast to New Mexico. Kyle drove half as far as Jamaal. Conner drove 4 times as far as Kyle. Altogether, they drove 476 miles. How far did each friend drive?

13. Bianca is taking collections for this year's Feed the Hungry Project. She has collected $300 more from Company A than from Company B and $700 more from Company C than from Company A. So far, she has collected $4, 300. How much did Company C give?

14. For his birthday, Torin got $50.00 more from his grandmother than from his uncle. His uncle gave him $10.00 less than his cousin. Torin received $135.00 in total. How much did he receive from his cousin?

15. Cassidy loves black and yellow jelly beans. She noticed when she was counting them that she had 7 less than three times as many black jelly beans as she had yellow jelly beans. In total, she counted 225 jelly beans. How many black jelly beans did she have?

16. Mrs. Vargus planted a garden with red and white rose bushes. Because she was studying to be a botanist, she counted the number of blossoms on each bush. She counted 4 times as many red blossoms as white blossoms. In total, she counted 1, 420 blossoms. How many red blossoms did she count?

Copyright © American Book Company

Chapter 10   Algebra Word Problems

10.5   Equivalent Forms of Equations

Example 8:   Harrison's Towing Company charges can be found using the equation
$y = 3x + 21$, where x is the number of miles a car is towed and y is the total
charges. Gene knows the amount he was charged when using Harrison's Towing
Company and wants to find the number of miles his car was towed. Find the
equation Gene should use to find the number of miles his car was towed.

Step 1:   The number of miles a car is towed equals x in the equation. Solve the original
equation for x.

Step 2:

$$
\begin{aligned}
y &= 3x + 21 \\
-21 & \qquad\quad -21 \\
\hline
\frac{y - 21}{3} &= \frac{3x}{3} \\
\tfrac{1}{3}y - 7 &= x
\end{aligned}
$$

The number of miles Gene's car was towed can be found with the equation
$x = \tfrac{1}{3}y - 7$.

Find the equations asked for in each problem.

1. A local bookstore is offering a special hard-cover edition of *Oliver Twist* for $25.95 minus
$0.10 for each used book the customer donates. The cost for the special hard-cover edition
of *Oliver Twist* can be expressed with the equation $g = 25.95 - 0.1a$, where a is the number
of books donated and g is the cost of the book. Find the equation you should use to find the
number of books donated.

2. Aja works for an air sanitizing company selling their products at a home improvement store.
Her salary for one week can be found using the equation $y = 20x + 480$, where y is her total
earnings for the week and x is the number of products she sold for the week. Aja knows how
much money she earned for the week and wants to find out how many products she sold. Find
the equation Aja should use to find the number of products she sold.

3. Branden had $800 in savings on January 1st. He deposits $50 every week. The amount of
money in Branden's account can be found using the equation $a = 50b + 800$, where a is the
total amount in his account and b is the number of weeks. Branden knows the total amount
in his account and wants to find the number of weeks that have gone by since he started
depositing money in his account each week. Find the equation Branden should use to find the
number of weeks.

4. The Rockbottom Blues band charges a $300 setup fee plus $175 per hour ($h$) that they play.
The total that they charge (c) can be found using the equation $c = 175h + 300$. Logan hired the
band for a party and knows how much they charged him. He wants to find the number of hours
they played. Find the equation Logan should use to find the number of hours the band played.

110   Copyright © American Book Company

10.6 Inequality Word Problems

10.6   Inequality Word Problems

Inequality word problems involve staying under a limit or having a minimum goal one must meet.

Example 9:   A contestant on a popular game show must earn a minimum of 800 points by answering a series of questions worth 40 points each per category in order to win the game. The contestant will answer questions from each of four categories. Her results for the first three categories are as follows: 160 points, 200 points, and 240 points. Write an inequality which describes how many points, (p), the contestant will need on the last category in order to win.

Step 1:   Add to find out how many points she already has. $160 + 200 + 240 = 600$

Step 2:   Subtract the points she already has from the minimum points she needs. $800 - 600 = 200$. She must get at least 200 points in the last category to win. If she gets more than 200 points, that is okay, too. To express the number of points she needs, use the following inequality statement:

$p \geq 200$ The points she needs must be greater than or equal to 200.

Solve each of the following problems using inequalities and graph your answer on a number line.

1. Stella wants to place her money in a high interest money market account. However, she needs at least $1,500 to open an account. Each month, she sets aside some of her earnings in a savings account. In January through June, she added the following amounts to her savings: $145, $203, $210, $120, $102, and $115. Write an inequality which describes the amount of money she can set aside in July to qualify for the money market account.

2. A high school band program will receive $2,000 for selling $12,000 worth of coupon books. Six band classes participate in the sales drive. Classes 1–5 collect the following amounts of money: $2,400, $2,800, $1,500, $2,320, and $2,550. Write an inequality which describes the amount of money the sixth class must collect so that the band will receive $2,000.

3. A small elevator has a maximum capacity of 1,200 pounds before the cable holding it in place snaps. Six people get on the elevator. Five of their weights follow: 120, 240, 150, 215, and 170. Write an inequality which describes the amount the sixth person can weigh without snapping the cable.

4. A small high school class of 9 students were told they would receive a pizza party if their class average was 90% or higher on the next exam. Students 1–8 scored the following on the exam: 84, 95, 99, 87, 92, 93, 100, and 98. Write an inequality which describes the score the ninth student must make for the class to qualify for the pizza party.

5. Raymond wants to spend his entire credit limit on his credit card. His credit limit is $3,000. He purchases items costing $750, $1,120, $42, $159, $8, and $71. Write an inequality which describes the amounts Raymond can put on his credit card for his next purchases.

Copyright © American Book Company

Chapter 10   Algebra Word Problems

Chapter 10 Review

1. Deanna is four more than three times older than Ted. The sum of their ages is 60. How old is Ted?

2. Ross is five years older than twice his sister Holly's age. The difference in their ages is 14 years. How old is Holly?

3. The sum of two numbers is 27. The larger number is 6 more than twice the smaller number. What are the numbers?

4. One number is 8 more than the other number. Twice the smaller number is 7 more than the larger number. What are the numbers?

5. The perimeter of a triangle is 48 inches. The second side is four inches longer than the first side. The third side is one inch longer than the second. Find the length of each side.

6. The perimeter of a rectangle is 292 feet. The length of the rectangle is 4 feet less than 5 times the width. What is the length and width of the rectangle?

7. Jesse and Larry entered a pie eating contest. Jesse ate 5 less than twice as many pies as Larry. They ate a total of 16 pies. How many pies did Larry eat?

8. Lena and Jodie are sisters, and together they have 56 bottles of nail polish. Lena bought 4 more than half the bottles. How many did Jodie buy?

9. Janet and Artie want to play tug-of-war. Artie pulls with 200 pounds of force while Janet pulls with 60 pounds of force. In order to make this a fair contest, Janet enlists the help of her friends Trudi, Sherri, and Bridget who pull with 20, 25, and 55 pounds respectively. Write an inequality describing the minimum amount Janet's fourth friend must pull to beat Artie.

10. Mr. Chan purchased 90 shares of stock for $0.50 last month, and the shares are now worth $3.80 each. Write an inequality which describes how much profit, p, Mr. Chan can make.

11. Erin is looking for a new job. During her interviews, Company A says pay is determined by the equation $y = 13x - 12$, where x is the number of hours worked. How much will Erin make if she can only work ten hours at this company?

12. Cameron lives in Woodstock, Georgia. In his research, he found that the population was $10,050$ in the year 2000 and in 2007, the population was $23,000$. Assuming this data is linear, what is the predicted population for 2014?

13. Margie is baking cookies for a bake sale. She bakes 2 dozen cookies every 30 minutes and already has 240 cookies. The number of cookies Margie has can be found by using the equation $c = \frac{4}{5}t + 240$, where c is the total number of cookies and t is the time in minutes. Margie counted the number of cookies she has and wants to find the time it took her to make the cookies. Find the equation she should use to find the time it took her to make the cookies.

Chapter 10 Test

1. Joe, Craig, and Dylan have a combined weight of 326 pounds. Craig weighs 40 pounds more than Joe. Dylan weighs 12 pounds more than Craig. How many pounds does Craig weigh?

   A. 130
   B. 118
   C. 97
   D. 78
   E. 52

2. Jim takes great pride in decorating his float for the homecoming parade for his high school. With the $5,000 he has to spend, Jim buys 5,000 carnations at $0.30 each, 4,000 tulips at $0.60 each, and 300 irises at $0.25 each. Write an inequality which describes how many roses, r, Jim can buy if roses cost $0.80 each.

   F. $r \leq 6,250$
   G. $r \geq 6,250$
   H. $r > 6,250$
   J. $r \geq 1,281$
   K. $r \leq 1,281$

3. Jack had $600 in savings on June 1st. He withdraws $20 every week. The amount of money in Jack's account can be found using the equation $a = 600 - 20w$, where a is the total amount in his account and w is the number of weeks. Jack knows the total amount in his account and wants to find the number of weeks that have gone by. Which equation should Jack use to find the number of weeks?

   A. $w = \frac{1}{20}a - 30$
   B. $w = -\frac{1}{20}a - 30$
   C. $w = -\frac{1}{20}a + 30$
   D. $w = -20a + 600$
   E. $w = 20a + 600$

4. Timothy bought a car for $5,500 in 1975. In 2000, he learned his car was worth $65,000. Based on a linear model, how much was Timothy's car worth in 2006?

   F. $79,280
   G. $95,000
   H. $81,660
   J. $72,340
   K. $80,250

5. Eric is eight less than three times Kim's age. The sum of their ages is 20. How old is Eric?

   A. 7
   B. 9
   C. 14
   D. 15
   E. 13

6. The perimeter of a rectangle is 26 feet. The width of the rectangle is three less than the length. What is the length of the rectangle?

   F. 8
   G. 7
   H. 6
   J. 5
   K. 9

7. At a concrete company, those who score 82% on a test, with a tolerance of 8%, earn a 5% raise. John scored 71%, Steve scored 76%, Catherine scored 82%, and Mike scored 91%. Which people will earn the 5% raise?

   A. only Mike
   B. John and Mike
   C. Steve, Catherine, and Mike
   D. Steve and Catherine
   E. only Catherine

Copyright © American Book Company

Chapter 11
Polynomials

Polynomials are algebraic expressions which include **monomials** containing one term, **binomials** which contain two terms, and **trinomials**, which contain three terms. Expressions with more than three terms are called **polynomials**. **Terms** are separated by plus and minus signs.

Examples

Monomials	Binomials	Trinomials	Polynomials
$4f$	$4t + 9$	$x^2 + 2x + 3$	$x^3 - 3x^2 + 3x - 9$
$3x^3$	$9 - 7g$	$5x^2 - 6x - 1$	$p^4 + 2p^3 + p^2 - 5 + p9$
$4g^2$	$5x^2 + 7x$	$y^4 + 15y^2 + 100$	
2	$6x^3 - 8x$		

11.1 Adding and Subtracting Monomials

Two **monomials** are added or subtracted as long as the **variable and its exponent** are the **same**. This is called combining like terms. Use the same rules you used for adding and subtracting integers.

Example 1:
$$4x + 5x = 9x \qquad \begin{array}{r} 3x^4 \\ -8x^4 \\ \hline -5x^4 \end{array} \qquad 2x^2 - 9x^2 = -7x^2 \qquad \begin{array}{r} 5y \\ +2y \\ \hline 7y \end{array} \qquad 6y^3 - 5y^3 = y^3$$

Remember: When the integer in front of the variable is "1", it is usually not written. $1x^2$ is the same as x^2, and $-1x$ is the same as $-x$.

Add or subtract the following monomials.

1. $2x^2 + 5x^2$
2. $5t + 8t$
3. $9y^3 - 2y^3$
4. $6g - 8g$
5. $7y^2 + 8y^2$

6. $s^5 + s^5$
7. $-2x - 4x$
8. $4w^2 - w^2$
9. $z^4 + 9z^4$
10. $-k + 2k$

11. $3x^2 - 5x^2$
12. $9t + 2t$
13. $-7v^3 + 10v^3$
14. $-2x^3 + x^3$
15. $10y^4 - 5y^4$

16. $\begin{array}{r} y^4 \\ +2y^4 \\ \hline \end{array}$

17. $\begin{array}{r} 4x^3 \\ -9x^3 \\ \hline \end{array}$

18. $\begin{array}{r} 8t^2 \\ +7t^2 \\ \hline \end{array}$

19. $\begin{array}{r} -2y \\ -4y \\ \hline \end{array}$

20. $\begin{array}{r} 5w^2 \\ +8w^2 \\ \hline \end{array}$

21. $\begin{array}{r} 11t^3 \\ -4t^3 \\ \hline \end{array}$

22. $\begin{array}{r} -5z \\ +9z \\ \hline \end{array}$

23. $\begin{array}{r} 4w^5 \\ +w^5 \\ \hline \end{array}$

24. $\begin{array}{r} 7t^3 \\ -6t^3 \\ \hline \end{array}$

25. $\begin{array}{r} 3x \\ +8x \\ \hline \end{array}$

11.2 Adding Polynomials

When adding **polynomials,** make sure the exponents and variables are the same on the terms you are combining. The easiest way is to put the terms in columns with **like exponents** under each other. Each column is added as a separate problem. Fill in the blank spots with zeros if it helps you keep the columns straight. You never carry to the next column when adding polynomials.

Example 2: Add $3x^2 + 14$ and $5x^2 + 2x$

$$\begin{array}{r} 3x^2 + 0x + 14 \\ (+)\, 5x^2 + 2x + 0 \\ \hline 8x^2 + 2x + 14 \end{array}$$

Example 3: $(4x^3 - 2x) + (-x^3 - 4)$

$$\begin{array}{r} 4x^3 - 2x + 0 \\ (+)\, -x^3 + 0x - 4 \\ \hline 3x^3 - 2x - 4 \end{array}$$

Add the following polynomials.

1. $y^2 + 3y + 2$ and $2y^2 + 4$

2. $(5y^2 + 4y - 6) + (2y^2 - 5y + 8)$

3. $5x^3 - 2x^2 + 4x - 1$ and $3x^2 - x + 2$

4. $-p + 4$ and $5p^2 - 2p + 2$

5. $(w - 2) + (w^2 + 2)$

6. $4t^2 - 5t - 7$ and $8t + 2$

7. $t^4 + t + 8$ and $2t^3 + 4t - 4$

8. $(3s^3 + s^2 - 2) + (-2s^3 + 4)$

9. $(-v^2 + 7v - 8) + (4v^3 - 6v + 4)$

10. $6m^2 - 2m + 10$ and $m^2 - m - 8$

11. $-x + 4$ and $3x^2 + x - 2$

12. $(8t^2 + 3t) + (-7t^2 - t + 4)$

13. $(3p^4 + 2p^2 - 1) + (-5p^2 - p + 8)$

14. $12s^3 + 9s^2 + 2s$ and $s^3 + s^2 + s$

15. $(-9b^2 + 7b + 2) + (-b^2 + 6b + 9)$

16. $15c^2 - 11c + 5$ and $-7c^2 + 3c - 9$

17. $5c^3 + 2c^2 + 3$ and $2c^3 + 4c^2 + 1$

18. $-14x^3 + 3x^2 + 15$ and $7x^3 - 12$

19. $(-x^2 + 2x - 4) + (3x^2 - 3)$

20. $(y^2 - 11y + 10) + (-13y^2 + 5y - 4)$

21. $3d^5 - 4d^3 + 7$ and $2d^4 - 2d^3 - 2$

22. $(6t^5 - t^3 + 17) + (4t^5 + 7t^3)$

23. $4p^2 - 8p + 9$ and $-p^2 - 3p - 5$

24. $20b^3 + 15b$ and $-4b^2 - 5b + 14$

25. $(-2w + 11) + (w^3 + w - 4)$

26. $(25z^2 + 13z + 8) + (z^2 - 2z - 10)$

Copyright © American Book Company

115

Chapter 11   Polynomials

11.3   Subtracting Polynomials

When you subtract polynomials, it is important to remember to change all the signs in the subtracted polynomial (the subtrahend) and then add.

Example 4: $(4y^2 + 8y + 9) - (2y^2 + 6y - 4)$

Step 1:   Copy the subtraction problem into vertical form. Make sure you line up the terms with like exponents under each other.

$$4y^2 + 8y + 9$$
$$\underline{(-)\, 2y^2 + 6y - 4}$$

Step 2:   Change the subtraction sign to addition and all the signs of the subtracted polynomial to the opposite sign.

$$4y^2 + 8y + 9$$
$$\underline{(+) - 2y^2 - 6y + 4}$$
$$2y^2 + 2y + 13$$

Subtract the following polynomials.

1. $(2x^2 + 5x + 2) - (x^2 + 3x + 1)$

2. $(8y - 4) - (4y + 3)$

3. $(11t^3 - 4t^2 + 3) - (-t^3 + 4t^2 - 5)$

4. $(-3w^2 + 9w - 5) - (-5w^2 - 5)$

5. $(6a^5 - a^3 + a) - (7a^5 + a^2 - 3a)$

6. $(14c^4 + 20c^2 + 10) - (7c^4 + 5c^2 + 12)$

7. $(5x^2 - 9x) - (-7x^2 + 4x + 8)$

8. $(12y^3 - 8y^2 - 10) - (3y^3 + y + 9)$

9. $(-3h^2 - 7h + 7) - (5h^2 + 4h + 10)$

10. $(10k^3 - 8) - (-4k^3 + k^2 + 5)$

11. $(x^2 - 5x + 9) - (6x^2 - 5x + 7)$

12. $(12p^2 + 4p) - (9p - 2)$

13. $(-2m - 8) - (6m + 2)$

14. $(13y^3 + 2y^2 - 8y) - (2y^3 + 4y^2 - 7y)$

15. $(7g + 3) - (g^2 + 4g - 5)$

16. $(-8w^3 + 4w) - (-10w^3 - 4w^2 - w)$

17. $(12x^3 + x^2 - 10) - (3x^3 + 2x^2 + 1)$

18. $(2a^2 + 2a + 2) - (-a^2 + 3a + 3)$

19. $(c + 19) - (3c^2 - 7c + 2)$

20. $(-6v^2 + 12v) - (3v^2 + 2v + 6)$

21. $(4b^3 + 3b^2 + 5) - (7b^3 - 8)$

22. $(15x^3 + 5x^2 - 4) - (4x^3 - 4x^2)$

23. $(8y^2 - 2) - (11y^2 - 2y - 3)$

24. $(-z^2 - 5z - 8) - (3z^2 - 5z + 5)$

116   Copyright © American Book Company

11.4 Multiplying Monomials

When two monomials have the **same variable**, you can multiply them. Then, add the **exponents** together. If the variable has no exponent, it is understood that the exponent is 1.

Example 5: $\qquad 4x^4 \times 3x^2 = 12x^6 \qquad\qquad 2y \times 5y^2 = 10y^3$

Multiply the following monomials.

1. $6a^1 \times 9a^5$

2. $2x^6 \times 5x^3$

3. $4y^3 \times 3y^2$

4. $10t^2 \times 2t^2$

5. $2p^5 \times 4p^2$

6. $9b^2 \times 8b^1$

7. $3c^3 \times 3c^3$

8. $2d^8 \times 9d^2$

9. $6k^3 \times 5k^2$

10. $7m^5 \times m^1$

11. $11z^1 \times 2z^7$

12. $3w^4 \times 6w^5$

13. $4x^4 \times 5x^3$

14. $5n^2 \times 3n^3$

15. $8w^7 \times w^1$

16. $10s^6 \times 5s^3$

17. $4d^5 \times 4d^5$

18. $5y^2 \times 8y^6$

19. $7t^{10} \times 3t^5$

20. $6p^8 \times 2p^3$

21. $x^3 \times 2x^3$

When problems include negative signs, follow the rules for multiplying integers.

22. $-7s^4 \times 5s^3$

23. $-6a \times -9a^5$

24. $4x \times -x$

25. $-3y^2 \times -y^3$

26. $-5b^2 \times 3b^5$

27. $9c^4 \times -2c$

28. $-4t^3 \times 8t^3$

29. $10d \times -8d^7$

30. $-3g^6 \times -2g^3$

31. $-7s^4 \times 7s^3$

32. $-d^3 \times -2d$

33. $11p \times -2p^5$

34. $-5x^7 \times -3x^3$

35. $8z^4 \times 7z^4$

36. $-4w \times -5w^8$

37. $-5y^4 \times 6y^2$

38. $9x^3 \times -7x^5$

39. $-a^4 \times -a$

40. $-7k^2 \times 3k$

41. $-15t^2 \times -t^4$

42. $3x^8 \times 9x^2$

Copyright © American Book Company

Chapter 11   Polynomials

11.5   Multiplying Monomials by Polynomials

In the chapter on solving multi-step equations, you learned to remove parentheses by multiplying the number outside the parentheses by each term inside the parentheses: $2(4x - 7) = 8x - 14$. Multiplying monomials by polynomials works the same way.

Example 6: $-5t(2t^2 - 7t + 9)$

Step 1:   Multiply $-5t \times 2t^2 = -10t^3$

Step 2:   Multiply $-5t \times -7t = 35t^2$

Step 3:   Multiply $-5t \times 9 = -45t$

Step 4:   Arrange the answers horizontally in order: $-10t^3 + 35t^2 - 45t$

Remove parentheses in the following problems.

1. $3x(3x^2 + 4x - 1)$

2. $4y(y^3 - 7)$

3. $7a^2(2a^2 + 3a + 2)$

4. $-5d^3(d^2 - 5d)$

5. $2w(-4w^2 + 3w - 8)$

6. $8p(p^3 - 6p + 5)$

7. $-9b^2(-2b + 5)$

8. $2t(t^2 - 4t - 10)$

9. $10c(4c^2 + 3c - 7)$

10. $6z(2z^4 - 5z^2 - 4)$

11. $-9t^2(3t^2 + 5t + 6)$

12. $c(-3c - 5)$

13. $3p(p^3 - p^2 - 9)$

14. $-k^2(2k + 4)$

15. $-3(4m^2 - 5m + 8)$

16. $6x(-7x^3 + 10)$

17. $-w(w^2 - 4w + 7)$

18. $2y(5y^2 - y)$

19. $3d(d^5 - 7d^3 + 4)$

20. $-5t(-4t^2 - 8t + 1)$

21. $7(2w^2 - 9w + 4)$

22. $3y^2(y^2 - 11)$

23. $v^2(v^2 + 3v + 3)$

24. $8x(2x^3 + 3x + 1)$

25. $-5d(4d^2 + 7d - 2)$

26. $-k^2(-3k + 6)$

27. $3x(-x^2 - 5x + 5)$

28. $4z(4z^4 - z - 7)$

29. $-5y(9y^3 - 3)$

30. $2b^2(7b^2 + 4b + 4)$

118   Copyright © American Book Company

11.6   Removing Parentheses and Simplifying

In the following problem, you must multiply each set of parentheses by the numbers and variables outside the parentheses, and then add the polynomials to simplify the expressions.

Example 7: $8x\left(2x^2 - 5x + 7\right) - 3x\left(4x^2 + 3x - 8\right)$

Step 1:   Multiply to remove the first set of parentheses.

$$8x\left(2x^2 - 5x + 7\right) = 16x^3 - 40x^2 + 56x$$

Step 2:   Multiply to remove the second set of parentheses.

$$-3x\left(4x^2 + 3x - 8\right) = -12x^3 - 9x^2 + 24x$$

Step 3:   Copy each polynomial in columns, making sure the terms with the same variable and exponent are under each other. Add to simplify.

$$
\begin{array}{r}
16x^3 - 40x^2 + 56x \\
(+) - 12x^3 - 9x^2 + 24x \\
\hline
4x^3 - 49x^2 + 80x
\end{array}
$$

Remove the parentheses and simplify the following problems.

1. $4t\left(t + 7\right) + 5t\left(2t^2 - 4t + 1\right)$

2. $-5y\left(3y^2 - 5y + 3\right) - 6y\left(y^2 - 4y - 4\right)$

3. $-3\left(3x^2 + 4x\right) + 5x\left(x^2 + 3x + 2\right)$

4. $2b\left(5b^2 - 8b - 1\right) - 3b\left(4b + 3\right)$

5. $8d^2\left(3d + 4\right) - 7d\left(3d^2 + 4d + 5\right)$

6. $5a\left(3a^2 + 3a + 1\right) - \left(-2a^2 + 5a - 4\right)$

7. $3m\left(m + 7\right) + 8\left(4m^2 + m + 4\right)$

8. $4c^2\left(-6c^2 - 3c + 2\right) - 7c\left(5c^3 + 2c\right)$

9. $-8w\left(-w + 1\right) - 4w\left(3w - 5\right)$

10. $6p\left(2p^2 - 4p - 6\right) + 3p\left(p^2 + 6p + 9\right)$

Copyright © American Book Company

Chapter 11   Polynomials

11.7   Multiplying Two Binomials Using the FOIL Method

When you multiply two binomials such as $(x + 6)(x - 5)$, you must multiply each term in the first binomial by each term in the second binomial. The easiest way is to use the **FOIL** method. If you can remember the word **FOIL**, it can help you keep order when you multiply. The "F" stands for **first**, "O" stands for **outside**, "I" stands for **inside**, and "L" stands for **last**.

F FIRST	**O** OUTSIDE	**I** INSIDE	**L** LAST
Multiply the **first** terms in each binomial	Next, multiply the **outside** terms.	Then, multiply the **inside** terms.	Last, multiply the **last** terms.
$(x + 6)(x - 5)$	$(x + 6)(x - 5)$	$(x + 6)(x - 5)$	$(x + 6)(x - 5)$
$x \times x = x^2$	$x \times -5 = -5x$	$6 \times x = 6x$	$6 \times -5 = -30$
x^2 $+$	$-5x$ $+$	$6x$ $+$	-30

Now just combine like terms, $6x - 5x = x$, and write your answer.

$(x + 6)(x - 5) = x^2 + x - 30.$

Note: It is customary for mathematicians to write polynomials in descending order. That means that the term with the highest exponent comes first in a polynomial. The next highest exponent is second, and so on. When you use the **FOIL** method, the terms will always be in the customary order. You just need to combine like terms and write your answer.

1. $(y - 7)(y + 3)$

2. $(2x + 4)(x + 9)$

3. $(4b - 3)(3b - 4)$

4. $(6g + 2)(g - 9)$

5. $(7k - 5)(-4k - 3)$

6. $(8v - 2)(3v + 4)$

7. $(10p + 2)(4p + 3)$

8. $(3h - 9)(-2h - 5)$

9. $(w - 4)(w - 7)$

10. $(6x + 1)(x - 2)$

11. $(5t + 3)(2t - 1)$

12. $(4y - 9)(4y + 9)$

13. $(a + 6)(3a + 5)$

14. $(3z - 8)(z - 4)$

15. $(5c + 2)(6c + 5)$

16. $(y + 3)(y - 3)$

17. $(2w - 5)(4w + 6)$

18. $(7x + 1)(x - 4)$

19. $(6t - 9)(4t - 4)$

20. $(5b + 6)(6b + 2)$

21. $(2z + 1)(10z + 4)$

22. $(11w - 8)(w + 3)$

23. $(5d - 9)(9d + 9)$

24. $(9g + 2)(g - 2)$

25. $(4p + 7)(2p + 3)$

26. $(m + 5)(m - 5)$

27. $(8b - 8)(2b - 1)$

28. $(z + 3)(3z + 5)$

29. $(7y - 5)(y - 3)$

30. $(9x + 5)(3x - 1)$

31. $(3t + 1)(t + 10)$

32. $(2w - 9)(8w + 7)$

33. $(8s - 2)(s + 4)$

34. $(4k - 1)(8k + 9)$

35. $(h + 12)(h - 2)$

36. $(3x + 7)(7x + 3)$

37. $(2v - 6)(2v + 6)$

38. $(2x + 8)(2x - 3)$

39. $(k - 1)(6k + 12)$

40. $(3w + 11)(2w + 2)$

41. $(8y - 10)(5y - 3)$

42. $(6d + 13)(d - 1)$

43. $(7h + 3)(2h + 4)$

44. $(5n + 9)(5n - 5)$

45. $(6z + 5)(z - 8)$

120   Copyright © American Book Company

11.8   Simplifying Expressions with Exponents

Example 8:   **Simplify** $(2a + 5)^2$
When you simplify an expression such as $(2a + 5)^2$, write
the expression as two binomials and use FOIL to simplify.
$(2a + 5)^2 = (2a + 5)(2a + 5)$
Using FOIL we have $4a^2 + 10a + 10a + 25 = 4a^2 + 20a + 25$

Example 9:   **Simplify** $4(3a + 2)^2$
Using order of operations, we must simplify the exponent first.

$4(3a + 2)^2$

$4(3a + 2)(3a + 2)$

$4(9a^2 + 6a + 6a + 4)$

$4(9a^2 + 12a + 4)$ Now multiply by 4.

$4(9a^2 + 12a + 4) = 36a^2 + 48a + 16$

Multiply the following binomials.

1. $(y + 3)^2$

2. $2(2x + 4)^2$

3. $6(4b - 3)^2$

4. $5(6g + 2)^2$

5. $(-4k - 3)^2$

6. $3(-2h - 5)^2$

7. $-2(8v - 2)^2$

8. $(10p + 2)^2$

9. $6(-2h - 5)^2$

10. $6(w - 7)^2$

11. $2(6x + 1)^2$

12. $(9x + 2)^2$

13. $(5t + 3)^2$

14. $3(4y - 9)^2$

15. $8(a + 6)^2$

16. $4(3z - 8)^2$

17. $3(5c + 2)^2$

18. $4(3t + 9)^2$

Copyright © American Book Company

Chapter 11   Polynomials

Chapter 11 Review

Simplify.

1. $3a^4 + 20a^4$

2. $(8x^4y^5)(20xy^7)$

3. $-6z^4(z+3)$

4. $(5b^4)(7b^3)$

5. $8x^4 - 20x^4$

6. $(7p-5)-(3p+4)$

7. $-7t(3t+20)^2$

8. $(3w^3y^4)(5wy^7)$

9. $3(4g+3)^2$

10. $25d^5 - 20d^5$

11. $(8w-5)(w-9)$

12. $27t^4 + 5t^4$

13. $(8c^5)(20c^4)$

14. $(20x+4)(x+7)$

15. $5y(5y^4 - 20y + 4)$

16. $(9a^5b)(4ab^3)(ab)$

17. $(7w^6)(20w^{20})$

18. $9x^3 + 24x^3$

19. $27p^7 - 22p^7$

20. $(3s^5t^4)(5st^3)$

21. $(5d+20)(4d+8)$

22. $5w(-3w^4 + 8w - 7)$

23. $45z^6 - 20z^6$

24. $-8y^3 - 9y^3$

25. $(8x^5)(8x^7)$

26. $28p^4 + 20p^4$

27. $(a^4v)(4av)(a^3v^6)$

28. $5(6y-7)^2$

29. $(3c^4)(6c^9)$

30. $(5x^7y^3)(4xy^3)$

31. Add $4x^4 + 20x$ and $7x^4 - 9x + 4$

32. $5t(6t^4 + 5t - 6) + 9t(3t+3)$

33. Subtract $y^4 + 5y - 6$ from $3y^4 + 8$

34. $4x(5x^4 + 6x - 3) + 5x(x+3)$

35. $(6t-5)-(6t^4 + t - 4)$

36. $(5x+6)+(8x^4 - 4x + 3)$

37. Subtract $7a - 4$ from $a + 20$

38. $(-4y+5)+(5y-6)$

39. $4t(t+6)-7t(4t+8)$

40. Add $3c - 5$ and $c^4 - 3c - 4$

41. $4b(b-5)-(b^4 + 4b + 2)$

42. $(6k^4 + 7k)+(k^4 + k + 20)$

43. $(q^4r^3)(3qr^4)(4q^5r)$

44. $(7df)(d^5f^4)(4df)$

45. $(8g^4h^3)(g^3h^6)(6gh^3)$

46. $(9v^4x^3)(3v^6x^4)(4v^5x^5)$

47. $(3n^4m^4)(20n^4m)(n^3m^8)$

48. $(22t^4a^4)(5t^3a^9)(4t^6a)$

49. $2b(b-4)-(b^2 + 2b + 1)$

122   Copyright © American Book Company

Chapter 11 Test

1. $2x^2 + 5x^2 =$

  A. $10x^4$
  B. $7x^4$
  C. $7x$
  D. $10x^2$
  E. $7x^2$

2. $-8m^3 + m^3 =$

  F. $-8m^6$
  G. $-8m^9$
  H. $-9m^6$
  J. $-7m^3$
  K. $9m^3$

3. $(6x^3 + x^2 - 5) + (-3x^3 - 2x^2 + 4) =$

  A. $3x^3 - x^2 - 1$
  B. $3x^3 - 3x^2 - 1$
  C. $3x^3 - 3x^2 - 9$
  D. $-3x^3 - 3x^2 - 1$
  E. $9x^3 - x^2 - 1$

4. $(-7c^2 + 5c + 3) + (-c^2 - 7c + 2) =$

  F. $-3x^3 - 3x^2 - 1$
  G. $-8c^2 - 2c + 5$
  H. $-6c^2 - 12c + 5$
  J. $-8c^2 - 12c + 5$
  K. $-6c^2 + 2c + 5$

5. $(5x^3 - 4x^2 + 5) - (-2x^3 - 3x^2) =$

  A. $3x^3 + x^2 + 5$
  B. $3x^3 - 7x^2 + 5$
  C. $7x^3 - x^2 + 5$
  D. $7x^3 - 7x^2 - 5$
  E. $7x^3 - 7x^2 + 5$

6. $(-z^3 - 4z^2 - 6) - (3z^3 - 6z + 5) =$

  F. $-4z^3 - 4z^2 + 6z - 11$
  G. $-2z^3 - 10z - 1$
  H. $-4z^3 - 10z^2 - 1$
  J. $-2z^2 + 2z - 11$
  K. $-3z^3 - 10z^2 - 1$

7. $(-7d^5)(-3d^2) =$

  A. $-21d^7$
  B. $21d^{10}$
  C. $21d^7$
  D. $-21d^{10}$
  E. $-10d^{10}$

8. $(-5c^3d)(3c^5d^3)(2cd^4) =$

  F. $30c^{15}d^8$
  G. $15c^8d^{12}$
  H. $-17c^{15}d^{12}$
  J. $-30c^9d^8$
  K. $-30c^8d^7$

9. $-11j^2 \times -j^4 =$

  A. $11j^6$
  B. $11j^8$
  C. $-11j^6$
  D. $-11j^8$
  E. $11j^2$

10. $-6m^2(7m^2 + 5m - 6) =$

  F. $-42m^2 + 30m^3 - 36$
  G. $-42m^4 - 30m^3 + 36m^2$
  H. $-13m^4 - m^2 + 36m^2$
  J. $42m^4 - 30m^3 - 36m^2$
  K. $42m^5 - 5m - 6$

Copyright © American Book Company

Chapter 11   Polynomials

11. $-h^2(-4h + 5) =$

   A. $-4h^3 - 5h^2$
   B. $4h^3 - 5h^2$
   C. $-5h^2 - 5h^2$
   D. $-5h^3 - 5h^2$
   E. $-3h^2 + 5$

12. $4m(m - 5) + 3m(2m^2 - 6m + 4) =$

   F. $6m^3 - 14m^2 - 8m$
   G. $-8m^2 - 8m - 1$
   H. $7m - 14m^2 - 1$
   J. $10m^2 - 26m - 20$
   K. $6m^3 + 14m^2 + 8m$

13. $2h(3h^2 - 5h - 2) + 4h(h^2 + 6h + 8) =$

   A. $6h^3 + 19h^2 + 28h$
   B. $-8m^2 - 8m - 1$
   C. $7m - 14m^2 - 1$
   D. $10h^3 + 14h^2 + 28h$
   E. $10h^3 - 14h^2 - 28h$

14. Multiply the following binomial and simplify. $(x - 3)(x + 3)$

   F. $x^2 - 3x + 3x - 9$
   G. $x^2 - 6x - 9$
   H. $x^2 + 9$
   J. $x^2 + 6x + 9$
   K. $x^2 - 9$

15. Multiply the following binomial and simplify. $(x + 9)(x + 1)$

   A. $x^2 + 10x + 9$
   B. $x^2 + 10x + 10$
   C. $x^2 + 9x + 9$
   D. $x^2 + 9x + x + 9$
   E. $x^2 + 8x + 10$

16. Multiply the following binomial and simplify. $(x - 2)^2$

   F. $x^2 - 4x - 4$
   G. $x^2 - 2x + 4$
   H. $x^2 - 2x - 4$
   J. $x^2 + 4x + 4$
   K. $x^2 - 4x + 4$

17. $(x + 4)^2 = ?$

   A. $x^2 + 4$
   B. $x^2 + 16$
   C. $x^2 + 16x + 8$
   D. $x^2 + 8x + 16$
   E. $x^2 + 8x + 4$

Chapter 12
Factoring

12.1 Finding the Greatest Common Factor of Polynomials

In a multiplication problem, the numbers multiplied together are called **factors**. The answer to a multiplication problem is a called the **product**.

In the multiplication problem $5 \times 4 = 20$, 5 and 4 are factors and 20 is the product.

If we reverse the problem, $20 = 5 \times 4$, we say we have **factored** 20 into 5×4.

In this chapter, we will factor **polynomials**.

Example 1:   Find the greatest common factor of $2y^3 + 6y^2$.

  Step 1:   Look at the whole numbers. The greatest common factor of 2 and 6 is 2. Factor the 2 out of each term.

$2\left(y^3 + 3y^2\right)$

  Step 2:   Look at the remaining terms, $y^3 + 3y^2$. What are the common factors of each term?

$$\begin{array}{ccccccc} y^3 & = & y & \times & \boxed{y & \times & y} \\ 3y^2 & = & 3 & \times & \boxed{y & \times & y} \end{array} \longleftarrow \text{common factors} = y^2$$

  Step 3:   Factor 2 and y^2 out of each term: $2y^2\left(y + 3\right)$

  Check: $2y^2\left(y + 3\right) = 2y^3 + 6y^2$

Copyright © American Book Company

125

Chapter 12   Factoring

Factor by finding the greatest common factor in each of the following.

1. $6x^4 + 18x^2$
2. $14y^3 + 7y$
3. $4b^5 + 12b^3$
4. $10a^3 + 5$
5. $2y^3 + 8y^2$

6. $6x^4 - 12x^2$
7. $18y^2 - 12y$
8. $15a^3 - 25a^2$
9. $4x^3 + 16x^2$
10. $6b^2 + 21b^5$

11. $27m^3 + 18m^4$
12. $100x^4 - 25x^3$
13. $4b^4 - 12b^3$
14. $18c^2 + 24c$
15. $20y^3 + 30y^5$

16. $16x^2 - 24x^5$
17. $15a^4 - 25a^2$
18. $24b^3 + 16b^6$
19. $36y^4 + 9y^2$
20. $42x^3 + 49x$

Factoring larger polynomials with 3 or 4 terms works the same way.

Example 2: $4x^5 + 16x^4 + 12x^3 + 8x^2$

Step 1:   Find the greatest common factor of the whole numbers. 4 can be divided evenly into 4, 16, 12, and 8; therefore, 4 is the greatest common factor.

Step 2:   Find the greatest common factor of the variables. x^5, x^4, x^3, and x^2 can be divided by x^2, the lowest power of x in each term.

$$4x^5 + 16x^4 + 12x^3 + 8x^2 = 4x^2 \left(x^3 + 4x^2 + 3x + 2\right)$$

Factor each of the following polynomials.

1. $5a^3 + 15a^2 + 20a$
2. $18y^4 + 6y^3 + 24y^2$
3. $12x^5 + 21x^3 + x^2$
4. $6b^4 + 3b^3 + 15b^2$
5. $14c^3 + 28c^2 + 7c$
6. $15b^4 - 5b^2 + 20b$
7. $t^3 + 3t^2 - 5t$
8. $8a^3 - 4a^2 + 12a$
9. $16b^5 - 12b^4 - 10b^2$
10. $20x^4 + 16x^3 - 24x^2 + 28x$
11. $40b^7 + 30b^5 - 50b^3$
12. $20y^4 - 15y^3 + 30y^2$

13. $4x^5 + 8x^4 + 12x^3 + 6x^2$
14. $16x^5 + 20x^4 - 12x^3 + 24x^2$
15. $18y^4 + 21y^3 - 9y^2$
16. $3n^5 + 9n^3 + 12n^2 + 15n$
17. $4d^6 - 8d^2 + 2d$
18. $10w^2 + 4w + 2$
19. $6t^3 - 3t^2 + 9t$
20. $25p^5 - 10p^3 - 5p^2$
21. $18x^4 + 9x^2 - 36x$
22. $6b^4 - 12b^2 - 6b$
23. $y^3 + 3y^2 - 9y$
24. $10x^5 - 2x^4 + 4x^2$

126   Copyright © American Book Company

12.1 Finding the Greatest Common Factor of Polynomials

Example 3: Find the greatest common factor of $4a^3b^2 - 6a^2b^2 + 2a^4b^3$

Step 1: The greatest common factor of the whole numbers is 2.

$$4a^3b^2 - 6a^2b^2 + 2a^4b^3 = 2\left(2a^3b^2 - 3a^2b^2 + a^4b^3\right)$$

Step 2: Find the lowest power of each variable that is in each term. Factor them out of each term. The lowest power of a is a^2. The lowest power of b is b^2.

$$4a^3b^2 - 6a^2b^2 + 2a^4b^3 = 2a^2b^2\left(2a - 3 + a^2b\right)$$

Factor each of the following polynomials.

1. $3a^2b^2 - 6a^3b^4 + 9a^2b^3$

2. $12x^4y^3 + 18x^3y^4 - 24x^3y^3$

3. $20x^2y - 25x^3y^3$

4. $12x^2y - 20x^2y^2 + 16xy^2$

5. $8a^3b + 12a^2b + 20a^2b^3$

6. $36c^4 + 42c^3 + 24c^2 - 18c$

7. $14m^3n^4 - 28m^3n^2 + 42m^2n^3$

8. $16x^4y^2 - 24x^3y^2 + 12x^2y^2 - 8xy^2$

9. $32c^3d^4 - 56c^2d^3 + 64c^3d^2$

10. $21a^4b^3 + 27a^2b^3 + 15a^3b^2$

11. $4w^3t^2 + 6w^2t - 8wt^2$

12. $5pw^3 - 2p^2q^2 - 9p^3q$

13. $49x^3t^3 + 7xt^2 - 14xt^3$

14. $9cd^4 - 3d^4 - 6c^2d^3$

15. $12a^2b^3 - 14ab + 10ab^2$

16. $25x^4 + 10x - 20x^2$

17. $bx^3 - b^2x^2 + b^3x$

18. $4k^3a^2 + 22ka + 16k^2a^2$

19. $33w^4y^2 - 9w^3y^2 + 24w^2y^2$

20. $18x^3 - 9x^5 + 27x^2$

Copyright © American Book Company

127

Chapter 12   Factoring

12.2   Finding the Numbers

The next kind of factoring we will do requires thinking of two numbers with a certain sum and a certain product.

Example 4:   Which two numbers have a sum of 8 and a product of 12? In other words, what pair of numbers would answer both equations?

$$\underline{\hspace{1cm}} + \underline{\hspace{1cm}} = 8 \quad \text{and} \quad \underline{\hspace{1cm}} \times \underline{\hspace{1cm}} = 12$$

You may think $4 + 4 = 8$, but 4×4 does not equal 12.
Or you may think $7 + 1 = 8$, but 7×1 does not equal 12.

$6 + 2 = 8$ and $6 \times 2 = 12$, so 6 and 2 are the pair of numbers that will work in both equations.

For each problem below, find one pair of numbers that will solve both equations.

1. $\underline{\hspace{1cm}} + \underline{\hspace{1cm}} = 14$ and $\underline{\hspace{1cm}} \times \underline{\hspace{1cm}} = 40$
2. $\underline{\hspace{1cm}} + \underline{\hspace{1cm}} = 10$ and $\underline{\hspace{1cm}} \times \underline{\hspace{1cm}} = 21$
3. $\underline{\hspace{1cm}} + \underline{\hspace{1cm}} = 18$ and $\underline{\hspace{1cm}} \times \underline{\hspace{1cm}} = 81$
4. $\underline{\hspace{1cm}} + \underline{\hspace{1cm}} = 12$ and $\underline{\hspace{1cm}} \times \underline{\hspace{1cm}} = 20$
5. $\underline{\hspace{1cm}} + \underline{\hspace{1cm}} = 7$ and $\underline{\hspace{1cm}} \times \underline{\hspace{1cm}} = 12$
6. $\underline{\hspace{1cm}} + \underline{\hspace{1cm}} = 8$ and $\underline{\hspace{1cm}} \times \underline{\hspace{1cm}} = 15$
7. $\underline{\hspace{1cm}} + \underline{\hspace{1cm}} = 10$ and $\underline{\hspace{1cm}} \times \underline{\hspace{1cm}} = 25$
8. $\underline{\hspace{1cm}} + \underline{\hspace{1cm}} = 14$ and $\underline{\hspace{1cm}} \times \underline{\hspace{1cm}} = 48$
9. $\underline{\hspace{1cm}} + \underline{\hspace{1cm}} = 12$ and $\underline{\hspace{1cm}} \times \underline{\hspace{1cm}} = 36$
10. $\underline{\hspace{1cm}} + \underline{\hspace{1cm}} = 17$ and $\underline{\hspace{1cm}} \times \underline{\hspace{1cm}} = 72$
11. $\underline{\hspace{1cm}} + \underline{\hspace{1cm}} = 15$ and $\underline{\hspace{1cm}} \times \underline{\hspace{1cm}} = 56$
12. $\underline{\hspace{1cm}} + \underline{\hspace{1cm}} = 9$ and $\underline{\hspace{1cm}} \times \underline{\hspace{1cm}} = 18$
13. $\underline{\hspace{1cm}} + \underline{\hspace{1cm}} = 13$ and $\underline{\hspace{1cm}} \times \underline{\hspace{1cm}} = 40$
14. $\underline{\hspace{1cm}} + \underline{\hspace{1cm}} = 16$ and $\underline{\hspace{1cm}} \times \underline{\hspace{1cm}} = 63$
15. $\underline{\hspace{1cm}} + \underline{\hspace{1cm}} = 10$ and $\underline{\hspace{1cm}} \times \underline{\hspace{1cm}} = 16$
16. $\underline{\hspace{1cm}} + \underline{\hspace{1cm}} = 8$ and $\underline{\hspace{1cm}} \times \underline{\hspace{1cm}} = 16$
17. $\underline{\hspace{1cm}} + \underline{\hspace{1cm}} = 9$ and $\underline{\hspace{1cm}} \times \underline{\hspace{1cm}} = 20$
18. $\underline{\hspace{1cm}} + \underline{\hspace{1cm}} = 13$ and $\underline{\hspace{1cm}} \times \underline{\hspace{1cm}} = 36$
19. $\underline{\hspace{1cm}} + \underline{\hspace{1cm}} = 15$ and $\underline{\hspace{1cm}} \times \underline{\hspace{1cm}} = 50$
20. $\underline{\hspace{1cm}} + \underline{\hspace{1cm}} = 11$ and $\underline{\hspace{1cm}} \times \underline{\hspace{1cm}} = 30$

Copyright © American Book Company

12.3 More Finding the Numbers

Now that you have mastered positive numbers, take up the challenge of finding pairs of negative numbers or pairs where one number is negative and one is positive.

Example 5: Which two numbers have a sum of -3 and a product of -40? In other words, what pair of numbers would answer both equations?

$$\underline{\hspace{1cm}} + \underline{\hspace{1cm}} = -3 \quad \text{and} \quad \underline{\hspace{1cm}} \times \underline{\hspace{1cm}} = -40$$

It is faster to look at the factors of 40 first. 8 and 5 and 10 and 4 are possibilities. 8 and 5 have a difference of 3, and in fact, $5 + (-8) = -3$ and $5 \times (-8) = -40$. This pair of numbers, 5 and -8, will satisfy both equations.

For each problem below, find one pair of numbers that will solve both equations.

1. $\underline{\hspace{1cm}} + \underline{\hspace{1cm}} = -2$ and $\underline{\hspace{1cm}} \times \underline{\hspace{1cm}} = -35$

2. $\underline{\hspace{1cm}} + \underline{\hspace{1cm}} = 4$ and $\underline{\hspace{1cm}} \times \underline{\hspace{1cm}} = -5$

3. $\underline{\hspace{1cm}} + \underline{\hspace{1cm}} = 4$ and $\underline{\hspace{1cm}} \times \underline{\hspace{1cm}} = -12$

4. $\underline{\hspace{1cm}} + \underline{\hspace{1cm}} = -6$ and $\underline{\hspace{1cm}} \times \underline{\hspace{1cm}} = 8$

5. $\underline{\hspace{1cm}} + \underline{\hspace{1cm}} = 3$ and $\underline{\hspace{1cm}} \times \underline{\hspace{1cm}} = -40$

6. $\underline{\hspace{1cm}} + \underline{\hspace{1cm}} = 10$ and $\underline{\hspace{1cm}} \times \underline{\hspace{1cm}} = -11$

7. $\underline{\hspace{1cm}} + \underline{\hspace{1cm}} = 6$ and $\underline{\hspace{1cm}} \times \underline{\hspace{1cm}} = -27$

8. $\underline{\hspace{1cm}} + \underline{\hspace{1cm}} = 8$ and $\underline{\hspace{1cm}} \times \underline{\hspace{1cm}} = -20$

9. $\underline{\hspace{1cm}} + \underline{\hspace{1cm}} = -5$ and $\underline{\hspace{1cm}} \times \underline{\hspace{1cm}} = -24$

10. $\underline{\hspace{1cm}} + \underline{\hspace{1cm}} = -3$ and $\underline{\hspace{1cm}} \times \underline{\hspace{1cm}} = -28$

11. $\underline{\hspace{1cm}} + \underline{\hspace{1cm}} = -2$ and $\underline{\hspace{1cm}} \times \underline{\hspace{1cm}} = -48$

12. $\underline{\hspace{1cm}} + \underline{\hspace{1cm}} = -1$ and $\underline{\hspace{1cm}} \times \underline{\hspace{1cm}} = -20$

13. $\underline{\hspace{1cm}} + \underline{\hspace{1cm}} = -3$ and $\underline{\hspace{1cm}} \times \underline{\hspace{1cm}} = 2$

14. $\underline{\hspace{1cm}} + \underline{\hspace{1cm}} = 1$ and $\underline{\hspace{1cm}} \times \underline{\hspace{1cm}} = -30$

15. $\underline{\hspace{1cm}} + \underline{\hspace{1cm}} = -7$ and $\underline{\hspace{1cm}} \times \underline{\hspace{1cm}} = 12$

16. $\underline{\hspace{1cm}} + \underline{\hspace{1cm}} = 6$ and $\underline{\hspace{1cm}} \times \underline{\hspace{1cm}} = -16$

17. $\underline{\hspace{1cm}} + \underline{\hspace{1cm}} = 5$ and $\underline{\hspace{1cm}} \times \underline{\hspace{1cm}} = -24$

18. $\underline{\hspace{1cm}} + \underline{\hspace{1cm}} = -4$ and $\underline{\hspace{1cm}} \times \underline{\hspace{1cm}} = 4$

19. $\underline{\hspace{1cm}} + \underline{\hspace{1cm}} = -1$ and $\underline{\hspace{1cm}} \times \underline{\hspace{1cm}} = -42$

20. $\underline{\hspace{1cm}} + \underline{\hspace{1cm}} = -6$ and $\underline{\hspace{1cm}} \times \underline{\hspace{1cm}} = 8$

Chapter 12   Factoring

12.4   Factoring Trinomials

In the chapter on polynomials, you multiplied binomials (two terms) together, and the answer was a trinomial (three terms).

For example, $(x + 6)(x - 5) = x^2 + x - 30$

Now, you need to practice factoring a trinomial into two binomials.

Example 6:   Factor $x^2 + 6x + 8$

Step 1:   When the trinomial is in descending order as in the example above, you need to find a pair of numbers whose sum equals the number in the second term, while their product equals the third term. In the above example, find the pair of numbers that has a sum of 6 and a product of 8.

$$\underline{\hspace{1cm}} + \underline{\hspace{1cm}} = 6 \quad \text{and} \quad \underline{\hspace{1cm}} \times \underline{\hspace{1cm}} = 8$$

The pair of numbers that satisfy both equations is 4 and 2.

Step 2:   Use the pair of numbers in the binomials.

The factors of $x^2 + 6x + 8$ are $(x + 4)(x + 2)$

Check: To check, use the FOIL method.
$(x + 4)(x + 2) = x^2 + 4x + 2x + 8 = x^2 + 6x + 8$

Notice, when the second term and the third term of the trinomial are both positive, both numbers in the solution are positive.

Example 7:   Factor $x^2 - x - 6$ Find the pair of numbers where:

the sum is -1 and the product is -6

$$\underline{\hspace{1cm}} + \underline{\hspace{1cm}} = -1 \quad \text{and} \quad \underline{\hspace{1cm}} \times \underline{\hspace{1cm}} = -6$$

The pair of numbers that satisfies both equations is 2 and -3.
The factors of $x^2 - x - 6$ are $(x + 2)(x - 3)$

Notice, if the second term and the third term are negative, one number in the solution pair is positive, and the other number is negative.

130   Copyright © American Book Company

12.4   Factoring Trinomials

Example 8:   Factor $x^2 - 7x + 12$ Find the pair of numbers where:

the sum is -7 and the product is 12

$$\underline{-3} + \underline{-4} = -7 \quad \text{and} \quad \underline{-3} \times \underline{-4} = 12$$

The pair of numbers that satisfies both equations is -3 and -4
The factors of $x^2 - 7x + 12$ are $(x - 3)(x - 4)$.

Notice, if the second term of a trinomial is negative and the third term is positive, both numbers in the solution are negative.

Find the factors of the following trinomials.

1. $x^2 - x - 2$

2. $y^2 + y - 6$

3. $w^2 + 3w - 4$

4. $t^2 + 5t + 6$

5. $x^2 + 2x - 8$

6. $k^2 - 4k + 3$

7. $t^2 + 3t - 10$

8. $x^2 - 3x - 4$

9. $y^2 - 5y + 6$

10. $y^2 + y - 20$

11. $a^2 - a - 6$

12. $b^2 - 4b - 5$

13. $c^2 - 5c - 14$

14. $c^2 - c - 12$

15. $d^2 + d - 6$

16. $x^2 - 3x - 28$

17. $y^2 + 3y - 18$

18. $a^2 - 9a + 20$

19. $b^2 - 2b - 15$

20. $c^2 + 7c - 8$

21. $t^2 - 11t + 30$

22. $w^2 + 13w + 36$

23. $m^2 - 2m - 48$

24. $y^2 + 14y + 49$

25. $x^2 + 7x + 10$

26. $a^2 - 7a + 6$

27. $d^2 - 6d - 27$

Copyright © American Book Company

Chapter 12   Factoring

12.5   More Factoring Trinomials

Sometimes a trinomial has a greatest common factor which must be factored out first.

Example 9:   Factor $4x^2 + 8x - 32$

Step 1:   Begin by factoring out the greatest common factor, 4.

$$4\left(x^2 + 2x - 8\right)$$

Step 2:   Factor by finding a pair of numbers whose sum is 2 and product is -8. 4 and -2 will work, so

$$4\left(x^2 + 2x - 8\right) = 4\left(x + 4\right)\left(x - 2\right)$$

Check: Multiply to check. $4\left(x + 4\right)\left(x - 2\right) = 4x^2 + 8x - 32$

Factor the following trinomials. Be sure to factor out the greatest common factor first.

1. $2x^2 + 6x + 4$

2. $3y^2 - 9y + 6$

3. $2a^2 + 2a - 12$

4. $4b^2 + 28b + 40$

5. $3y^2 - 6y - 9$

6. $10x^2 + 10x - 200$

7. $5c^2 - 10c - 40$

8. $6d^2 + 30d - 36$

9. $4x^2 + 8x - 60$

10. $6a^2 - 18a - 24$

11. $5b^2 + 40b + 75$

12. $3c^2 - 6c - 24$

13. $2x^2 - 18x + 28$

14. $4y^2 - 20y + 16$

15. $7a^2 - 7a - 42$

16. $6b^2 - 18b - 60$

17. $11d^2 + 66d + 88$

18. $3x^2 - 24x + 45$

132   Copyright © American Book Company

12.6 Factoring More Trinomials

Some trinomials have a whole number in front of the first term that cannot be factored out of the trinomial. The trinomial can still be factored.

Example 10:   Factor $2x^2 + 5x - 3$

Step 1:   To get a product of $2x^2$, one factor must begin with $2x$ and the other with x.

$$(2x \quad)(x \quad)$$

Step 2:   Now think: What two numbers give a product of -3? The two possibilities are 3 and -1 or -3 and 1. We know they could be in any order so there are 4 possible arrangements.

$$(2x + 3)(x - 1)$$
$$(2x - 3)(x + 1)$$
$$(2x + 1)(x - 3)$$
$$(2x - 1)(x + 3)$$

Step 3:   Multiply each possible answer until you find the arrangement of the numbers that works. Multiply the outside terms and the inside terms and add them together to see which one will equal $5x$.

$$(2x + 3)(x - 1) = 2x^2 + x - 3$$
$$(2x - 3)(x + 1) = 2x^2 - x - 3$$
$$(2x + 1)(x - 3) = 2x^2 - 5 - 3$$
$$\boxed{(2x - 1)(x + 3) = 2x^2 + 5x - 3} \longleftarrow \text{ This arrangement works, therefore:}$$

The factors of $2x^2 + 5x - 3$ are $(2x - 1)(x + 3)$

Alternative: You can do some of the multiplying in your head. For the above example, ask yourself the following question: What two numbers give a product of -3 and give a sum of 5 (the whole number in the second term) when one number is first multiplied by 2 (the whole number in front of the first term)? The pair of numbers, -1 and 3, have a product of -3 and a sum of 5 when the 3 is first multiplied by 2. Therefore, the 3 will go in the opposite factor of the $2x$ so that when the terms are multiplied, you get -5.

You can use this method to at least narrow down the possible pairs of numbers when you have several from which to choose.

Copyright © American Book Company

Chapter 12   Factoring

Factor the following trinomials.

1. $3y^2 + 14y + 8$

2. $5a^2 + 24a - 5$

3. $7b^2 + 30b + 8$

4. $2c^2 - 9c + 9$

5. $2y^2 - 7y - 15$

6. $3x^2 + 4x + 1$

7. $7y^2 + 13y - 2$

8. $11a^2 + 35a + 6$

9. $5y^2 + 17y - 12$

10. $3a^2 + 4a - 7$

11. $2a^2 + 3a - 20$

12. $5b^2 - 13b - 6$

13. $3y^2 - 4y - 32$

14. $2x^2 - 17x + 36$

15. $11x^2 - 29x - 12$

16. $5c^2 + 2c - 16$

17. $7y^2 - 30y + 27$

18. $2x^2 - 3x - 20$

19. $5b^2 - 19b - 4$

20. $7d^2 + 18d + 8$

21. $3x^2 - 20x + 25$

22. $2a^2 - 7a - 4$

23. $5m^2 + 12m + 4$

24. $9y^2 - 5y - 4$

25. $2b^2 - 13b + 18$

26. $7x^2 + 31x - 20$

27. $3c^2 - 2c - 21$

134   Copyright © American Book Company

12.7 Factoring the Difference of Two Squares

Let's give an example of a **perfect square**.

25 is a perfect square because $5 \times 5 = 25$
49 is a perfect square because $7 \times 7 = 49$

Any variable with an even exponent is a perfect square.

y^2 is a perfect square because $y \times y = y^2$
y^4 is a perfect square because $y^2 \times y^2 = y^4$

When two terms that are both perfect squares are subtracted, factoring those terms is very easy. To factor the difference of perfect squares, you use the square root of each term, a plus sign in the first factor, and a minus sign in the second factor.

Example 11: Factor $4x^2 - 9$

This example has two terms which are both perfect squares, and the terms are subtracted.

Step 1:

Find the square root of each term.
Use the square roots in each of the factors.

Step 2: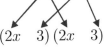

Use a plus sign in one factor
and a minus sign in the other factor.

Check: Multiply to check. $(2x + 3)(2x - 3) = 4x^2 - 6x + 6x - 9 = 4x^2 - 9$

The inner and outer terms add to zero.

Example 12: Factor $81y^4 - 1$

Step 1: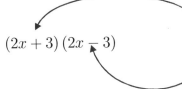

Factor like the example above.
Notice, the second factor is also the difference of two perfect squares.

Step 2: $(9y^2 + 1)(3y + 1)(3y - 1)$

Factor the second term further.
Note: You cannot factor the sum of two perfect squares.

Check: Multiply in reverse to check your answer.
$(9y^2 + 1)(3y + 1)(3y - 1) = (9y^2 + 1)(9y^2 - 3y + 3y - 1) =$
$(9y^2 + 1)(9y^2 - 1) = 81y^4 + 9y^2 - 9y^2 - 1 = 81y^4 - 1$

Chapter 12   Factoring

Factor the following differences of perfect squares.

1. $64x^2 - 49$

2. $4y^4 - 25$

3. $9a^4 - 4$

4. $25c^4 - 9$

5. $64y^2 - 9$

6. $x^4 - 16$

7. $49x^2 - 4$

8. $4d^2 - 25$

9. $9a^2 - 16$

10. $100y^4 - 49$

11. $c^4 - 36$

12. $36x^2 - 25$

13. $25x^2 - 4$

14. $9x^4 - 64$

15. $49x^2 - 100$

16. $16x^2 - 81$

17. $9y^4 - 1$

18. $49c^2 - 25$

19. $25d^2 - 64$

20. $36a^4 - 49$

21. $16x^4 - 16$

22. $b^2 - 25$

23. $c^4 - 144$

24. $9y^2 - 4$

25. $81x^4 - 16$

26. $4b^2 - 36$

27. $9w^2 - 9$

28. $64a^2 - 25$

29. $49y^2 - 121$

30. $x^6 - 9$

Chapter 12 Review

Factor the following polynomials completely.

1. $8x - 18$

2. $6x^2 - 18x$

3. $16b^3 + 8b$

4. $15a^3 + 40$

5. $20y^6 - 12y^4$

6. $5a - 15a^2$

7. $4y^2 - 36$

8. $2b^2 - 2b - 12$

9. $27y^2 + 42y - 5$

10. $12b^2 + 25b - 7$

11. $c^2 + cd - 20d^2$

12. $6y^2 + 30y + 36$

13. $2b^2 + 6b - 20$

14. $16b^4 - 81d^4$

15. $9w^2 - 54w - 63$

16. $12x^2 + 27x$

17. $2a^4 - 32$

18. $21c^2 + 41c + 10$

Copyright © American Book Company

137

Chapter 12 Factoring

Chapter 12 Test

1. What is the greatest common factor of $4x^3$ and $8x^2$?

   A. $4x^2$
   B. $4x$
   C. x^2
   D. $8x$
   E. $32x^5$

2. Factor: $8x^4 - 7x^2 + 4x$

   F. $4x\left(2x^3 - 7x + 4\right)$
   G. $x\left(8x^4 - 7x^2 + 4x\right)$
   H. $x\left(8x^3 - 7x + 4\right)$
   J. $4x\left(2x^3 - 7x + 1\right)$
   K. $x(8^4 - 7x + 4)$

3. Factor: $x^2 + 6x + 8$

   A. $(x + 2)\left(x + 2\right)$
   B. $(x + 1)\left(x + 8\right)$
   C. $(x - 2)\left(x - 4\right)$
   D. $(x - 1)\left(x - 8\right)$
   E. $(x + 2)\left(x + 4\right)$

4. Factor: $2x^2 - 2x - 84$

   F. $(2x + 7)\left(x - 12\right)$
   G. $(2x - 12)\left(x + 7\right)$
   H. $(2x - 7)\left(x + 12\right)$
   J. $(2x + 12)\left(x - 7\right)$
   K. $(2x - 7)\left(x - 12\right)$

5. Factor: $4x^2 - 64$

   A. $(x - 8)\left(x + 8\right)$
   B. $(4x - 8)\left(4x + 8\right)$
   C. $(2x - 16)\left(2x + 16\right)$
   D. $(2x - 8)\left(2x - 8\right)$
   E. $(2x - 8)\left(2x + 8\right)$

6. Factor the greatest common factor out of $2x^3 - 6x^2 + 2x$.

   F. $2x\left(x^2 - 6x + 1\right)$
   G. $2x\left(x^2 - 3x + 1\right)$
   H. $2x\left(x^2 - 3x\right)$
   J. $2\left(x^3 - 3x^2 + x\right)$
   K. $2x\left(x^2 - 2x + 1\right)$

7. What are the factors of $x^2 + 10x + 25$?

   A. $(x + 5)^2$
   B. $(x - 5)^2$
   C. $(x - 5)(x + 5)$
   D. $(x + 5)(x + 2)$
   E. $(x + 5)(x - 1)$

8. What are the factors of $x^2 + 11x + 30$?

   F. $(x + 6)(x + 5)$
   G. $(x + 10)(x + 3)$
   H. $(x + 6)^2$
   J. $(x + 15)(x + 2)$
   K. $(x + 10)(x + 1)$

9. Factor: $3x^2 - 9$

   A. $(3x + 3)(x - 3)$
   B. $(x + 3)(x - 3)$
   C. $3\left(x^2 - 3\right)$
   D. $(3x + 9)(x - 1)$
   E. $(x + 3)(2x - 3)$

10. Factor $4y^4 - 36$

   F. $(2y^2 + 6)(2y^2 - 6)$
   G. $(2y^2 + 6)(2y^2 + 6)$
   H. $(2y^2 + 6)^2$
   J. $(2y^2 + 6)(2y^2 - 30)$
   K. $(2y^2 + 5)(2y^2 - 7)$

Copyright © American Book Company

Chapter 13
Solving Quadratic Equations

13.1   Solving Quadratic Equations

In the previous chapter, we factored polynomials such as $y^2 - 4y - 5$ into two factors:

$$y^2 - 4y - 5 = (y + 1)(y - 5)$$

In this chapter, we learn that any equation that can be put in the form $ax^2 + bx + c = 0$ is a quadratic equation if a, b, and c are real numbers and $a \neq 0$. $ax^2 + bx + c = 0$ is the standard form of a quadratic equation. To solve these equations, follow the steps below.

Example 1:   Solve $y^2 - 4y - 5 = 0$

Step 1:   Factor the left side of the equation.

$$\begin{aligned} y^2 - 4y - 5 &= 0 \\ (y+1)(y-5) &= 0 \end{aligned}$$

Step 2:   If the product of these two factors equals zero, then the two factors individually must be equal to zero. Therefore, to solve, we set each factor equal to zero.

$$\begin{array}{r} (y+1) = 0 \\ \underline{-1 \quad -1} \\ y = -1 \end{array} \qquad\qquad \begin{array}{r} (y-5) = 0 \\ \underline{+5 \quad +5} \\ y = 5 \end{array}$$

The equation has two solutions: $y = -1$ and $y = 5$

Check: To check, substitute each solution into the original equation.

When $y = -1$, the equation becomes:
$$\begin{aligned} (-1)^2 - (4)(-1) - 5 &= 0 \\ 1 + 4 - 5 &= 0 \\ 0 &= 0 \end{aligned}$$

When $y = 5$, the equation becomes:
$$\begin{aligned} 5^2 - (4)(5) - 5 &= 0 \\ 25 - 20 - 5 &= 0 \\ 0 &= 0 \end{aligned}$$

Both solutions produce true statements.
The solution set for the equation is $\{-1, 5\}$.

Copyright © American Book Company   139

Chapter 13   Solving Quadratic Equations

Solve each of the following quadratic equations by factoring and setting each factor equal to zero. Check by substituting answers back in the original equation.

1. $x^2 + x - 6 = 0$

2. $y^2 - 2y - 8 = 0$

3. $a^2 + 2a - 15 = 0$

4. $y^2 - 5y + 4 = 0$

5. $b^2 - 9b + 14 = 0$

6. $x^2 - 3x - 4 = 0$

7. $y^2 + y - 20 = 0$

8. $d^2 + 6d + 8 = 0$

9. $y^2 - 7y + 12 = 0$

10. $x^2 - 3x - 28 = 0$

11. $a^2 - 5a + 6 = 0$

12. $b^2 + 3b - 10 = 0$

13. $a^2 + 7a - 8 = 0$

14. $c^2 + 3x + 2 = 0$

15. $x^2 - x - 42 = 0$

16. $a^2 + a - 6 = 0$

17. $b^2 + 7b + 12 = 0$

18. $y^2 + 2y - 15 = 0$

19. $a^2 - 3a - 10 = 0$

20. $d^2 + 10d + 16 = 0$

21. $x^2 - 4x - 12 = 0$

Quadratic equations that have a whole number and a variable in the first term are solved the same way as the previous page. Factor the trinomial, and set each factor equal to zero to find the solution set.

Example 2:   Solve $2x^2 + 3x - 2 = 0$
$(2x - 1)(x + 2) = 0$
Set each factor equal to zero and solve:

$$\begin{array}{r} 2x - 1 = 0 \\ \underline{+1 \quad +1} \\ \dfrac{2x}{2} = \dfrac{1}{2} \\ x = \dfrac{1}{2} \end{array} \qquad \begin{array}{r} x + 2 = 0 \\ \underline{-2 \quad -2} \\ x = -2 \end{array}$$

The solution set is $\left\{ \dfrac{1}{2}, -2 \right\}$.

Solve the following quadratic equations.

22. $3y^2 + 4y - 32 = 0$

23. $5c^2 - 2c - 16 = 0$

24. $7d^2 + 18d + 8 = 0$

25. $3a^2 - 10a - 8 = 0$

26. $11x^2 - 31x - 6 = 0$

27. $5b^2 + 17b + 6 = 0$

28. $3x^2 - 11x - 20 = 0$

29. $5a^2 + 47a - 30 = 0$

30. $2c^2 - 5c - 25 = 0$

31. $2y^2 + 11y - 21 = 0$

32. $5a^2 + 23a - 42 = 0$

33. $3d^2 + 11d - 20 = 0$

34. $3x^2 - 10x + 8 = 0$

35. $7b^2 + 23b - 20 = 0$

36. $9a^2 - 58a + 24 = 0$

37. $4c^2 - 25c - 21 = 0$

38. $8d^2 + 53d + 30 = 0$

39. $4y^2 - 29y + 30 = 0$

40. $8a^2 + 37a - 15 = 0$

41. $3x^2 - 41x + 26 = 0$

42. $8b^2 + 2b - 3 = 0$

13.2 Solving the Difference of Two Squares

To solve the difference of two squares, first factor. Then set each factor equal to zero.

Example 3: $25x^2 - 36 = 0$

Step 1: Factor the left side of the equation.

$$25x^2 - 36 = 0$$
$$(5x + 6)(5x - 6) = 0$$

Step 2: Set each factor equal to zero and solve.

$$
\begin{array}{rcl}
5x + 6 & = & 0 \\
-6 & & -6 \\
\hline
\dfrac{5x}{5} & = & \dfrac{6}{5} \\
x & = & -\dfrac{6}{5}
\end{array}
\qquad\qquad
\begin{array}{rcl}
5x - 6 & = & 0 \\
+6 & & +6 \\
\hline
\dfrac{5x}{5} & = & \dfrac{6}{5} \\
x & = & \dfrac{6}{5}
\end{array}
$$

Check: Substitute each solution in the equation to check.

for $x = -\dfrac{6}{5}$:

$$25x^2 - 36 = 0$$

$$25\left(-\frac{6}{5}\right)\left(-\frac{6}{5}\right) - 36 = 0 \longleftarrow \text{ Substitute } -\tfrac{6}{5} \text{ for } x.$$

$$25\left(\frac{36}{25}\right) - 36 = 0 \longleftarrow \text{ Cancel the 25's.}$$

$$36 - 36 = 0 \longleftarrow \text{ A true statement. } x = -\tfrac{6}{5} \text{ is a solution.}$$

for $x = \dfrac{6}{5}$:

$$25x^2 - 36 = 0$$

$$25\left(\frac{6}{5}\right)\left(\frac{6}{5}\right) - 36 = 0 \longleftarrow \text{ Substitute } \tfrac{6}{5} \text{ for } x.$$

$$25\left(\frac{36}{25}\right) - 36 = 0 \longleftarrow \text{ Cancel the 25's.}$$

$$36 - 36 = 0 \longleftarrow \text{ A true statement. } x = \tfrac{6}{5} \text{ is a solution.}$$

The solution set is $\left\{-\dfrac{6}{5}, \dfrac{6}{5}\right\}$.

Copyright © American Book Company

Chapter 13   Solving Quadratic Equations

Find the solution sets for the following.

1. $25a^2 - 16 = 0$

2. $c^2 - 36 = 0$

3. $9x^2 - 64 = 0$

4. $100y^2 - 49 - 0$

5. $4b^2 - 81 = 0$

6. $d^2 - 25 = 0$

7. $9x^2 - 1 = 0$

8. $16a^2 - 9 = 0$

9. $36y^2 - 1 = 0$

10. $36y^2 - 25 = 0$

11. $d^2 - 16 = 0$

12. $64b^2 - 9 = 0$

13. $81a^2 - 4 = 0$

14. $64y^2 - 25 = 0$

15. $4c^2 - 49 = 0$

16. $x^2 - 81 = 0$

17. $49b^2 - 9 = 0$

18. $a^2 - 64 = 0$

19. $9x^2 - 1 = 0$

20. $4y^2 - 9 = 0$

21. $t^2 - 100 = 0$

22. $16k^2 - 81 = 0$

23. $81a^2 - 4 = 0$

24. $36b^2 - 16 = 0$

13.3 Solving Perfect Squares

13.3 Solving Perfect Squares

When the square root of a constant, variable, or polynomial results in a constant, variable, or polynomial without irrational numbers, the expression is a **perfect square**. Some examples are 49, x^2, and $(x-2)^2$.

Example 4: Solve the perfect square for x. $(x-5)^2 = 0$

Step 1: Take the square root of both sides.
$$\sqrt{(x-5)^2} = \sqrt{0}$$
$$(x-5) = 0$$

Step 2: Solve the equation.
$$(x-5) = 0$$
$$x - 5 + 5 = 0 + 5$$
$$x = 5$$

Example 5: Solve the perfect square for x. $(x-5)^2 = 64$

Step 1: Take the square root of both sides.
$$\sqrt{(x-5)^2} = \sqrt{64}$$
$$(x-5) = \pm 8$$
$$(x-5) = 8 \text{ and } (x-5) = -8$$

Step 2: Solve the two equations.
$$\begin{array}{lll} (x-5) = 8 & \text{and} & (x-5) = -8 \\ x - 5 + 5 = 8 + 5 & \text{and} & x - 5 + 5 = -8 + 5 \\ x = 13 & \text{and} & x = -3 \end{array}$$

Solve the perfect square for x.

1. $(x-5)^2 = 0$

2. $(x+1)^2 = 0$

3. $(x+11)^2 = 0$

4. $(x-4)^2 = 0$

5. $(x-1)^2 = 0$

6. $(x+8)^2 = 0$

7. $(x+3)^2 = 4$

8. $(x-5)^2 = 16$

9. $(x-10)^2 = 100$

10. $(x+9)^2 = 9$

11. $(x-4.5)^2 = 25$

12. $(x+7)^2 = 36$

13. $(x+2)^2 = 49$

14. $(x-1)^2 = 4$

15. $(x+8.9)^2 = 49$

16. $(x-6)^2 = 81$

17. $(x-12)^2 = 121$

18. $(x+2.5)^2 = 64$

Copyright © American Book Company

Chapter 13   Solving Quadratic Equations

13.4   Using the Quadratic Formula

You may be asked to use the quadratic formula to solve an algebra problem known as a **quadratic equation**. The equation should be in the form $ax^2 + bx + c = 0$.

Example 6:   Using the quadratic formula, find x in the following equation: $x^2 - 8x = -7$.

Step 1:   Make sure the equation is set equal to 0.

$$x^2 - 8x + 7 = -7 + 7$$
$$x^2 - 8x + 7 = 0$$

The quadratic formula is $\dfrac{-b \pm \sqrt{b^2 - 4ac}}{2a}$.

Step 2:   In the formula, a is the number x^2 is multiplied by, b is the number x is multiplied by and c is the last term of the equation. For the equation in the example, $x^2 - 8x + 7$, $a = 1$, $b = -8$, and $c = 7$. When we look at the formula we notice a $\pm$ sign. This means that there will be two solutions to the equation, one when we use the plus sign and one when we use the minus sign. Substituting the numbers from the problem into the formula, we have:

$$\frac{8 + \sqrt{(-8)^2 - (4)(1)(7)}}{2(1)} = 7 \qquad \text{or} \qquad \frac{8 - \sqrt{(-8)^2 - (4)(1)(7)}}{2(1)} = 1$$

The solutions are $\{1, 7\}$.

For each of the following equations, use the quadratic formula to find two solutions.

1. $x^2 + x - 6 = 0$

2. $y^2 - 2y - 8 = 0$

3. $a^2 + 2a - 15 = 0$

4. $y^2 - 5y + 4 = 0$

5. $b^2 - 9b + 14 = 0$

6. $x^2 - 3x - 4 = 0$

7. $y^2 + y - 20 = 0$

8. $d^2 + 6d + 8 = 0$

9. $y^2 - 7y + 12 = 0$

10. $x^2 - 3x - 28 = 0$

11. $a^2 - 5a + 6 = 0$

12. $b^2 + 3b - 10 = 0$

13. $a^2 + 7a - 8 = 0$

14. $c^2 + 3c + 2 = 0$

15. $x^2 - x - 42 = 0$

16. $a^2 + 5a - 6 = 0$

17. $b^2 + 7b + 12 = 0$

18. $y^2 + y - 12 = 0$

19. $a^2 - 3a - 10 = 0$

20. $d^2 + 10d + 16 = 0$

21. $x^2 - 4x - 12 = 0$

144   Copyright © American Book Company

Chapter 13 Review

Factor and solve each of the following quadratic equations.

1. $16b^2 - 25 = 0$

2. $a^2 - a - 30 = 0$

3. $x^2 - x = 6$

4. $100x^2 - 49 = 0$

5. $81y^2 = 9$

6. $y^2 = 21 - 4y$

7. $y^2 - 7y + 8 = 16$

8. $6x^2 + x - 2 = 0$

9. $3y^2 + y - 2 = 0$

10. $b^2 + 2b - 8 = 0$

11. $4x^2 + 19x - 5 = 0$

12. $8x^2 = 6x + 2$

13. $2y^2 - 6y - 20 = 0$

14. $-6x^2 + 7x - 2 = 0$

15. $y^2 + 3y - 18 = 0$

Using the quadratic formula, find both solutions for the variable.

16. $x^2 + 10x - 11 = 0$

17. $y^2 - 14y + 40 = 0$

18. $b^2 + 9b + 18 = 0$

19. $y^2 - 12y - 13 = 0$

20. $a^2 - 8a - 48 = 0$

21. $x^2 + 2x - 63 = 0$

Chapter 13   Solving Quadratic Equations

Chapter 13 Test

1. Solve: $4y^2 - 9y = -5$

   A. $\left\{1, \dfrac{5}{4}\right\}$

   B. $\left\{-\dfrac{3}{4}, -1\right\}$

   C. $\left\{-1, \dfrac{4}{5}\right\}$

   D. $\left\{\dfrac{5}{16}, 1\right\}$

   E. $\left\{1, -\dfrac{5}{4}\right\}$

2. Solve for y: $2y^2 + 13y + 15 = 0$

   F. $\left\{\dfrac{3}{2}, \dfrac{5}{2}\right\}$

   G. $\left\{\dfrac{2}{3}, \dfrac{2}{5}\right\}$

   H. $\left\{-5, -\dfrac{3}{2}\right\}$

   J. $\left\{5, -\dfrac{3}{2}\right\}$

   K. $\left\{\dfrac{2}{3}, -\dfrac{2}{5}\right\}$

3. Solve for x.

   $x^2 - 3x - 18 = 0$

   A. $\{-6, 3\}$
   B. $\{6, -3\}$
   C. $\{-9, 2\}$
   D. $\{9, -2\}$
   E. $\{6, -2\}$

4. What are the values of x in the quadratic equation?

   $x^2 + 2x - 15 = x - 3$

   F. $\{4, 3\}$
   G. $\{-3, 4\}$
   H. $\{-3, 5\}$
   J. $\{-4, -3\}$
   K. $\{-4, 3\}$

5. Solve the equation $(x + 9)^2 = 49$

   A. $x = -9, 9$
   B. $x = -9, 7$
   C. $x = -16, -2$
   D. $x = -7, 7$
   E. $x = -9, 2$

6. Solve: $x = \sqrt{\dfrac{10 - 8x}{2}}$

   F. $\{0, 6\}$
   G. $\{-5, 10\}$
   H. $\{4, -2\}$
   J. $\{-5, 1\}$
   K. $\{-4, 10\}$

7. Solve $6a^2 + 11a - 10 = 0$, using the quadratic formula.

   A. $\left\{-\dfrac{2}{5}, \dfrac{3}{2}\right\}$

   B. $\left\{\dfrac{2}{5}, \dfrac{2}{3}\right\}$

   C. $\left\{-\dfrac{5}{2}, \dfrac{2}{3}\right\}$

   D. $\left\{\dfrac{5}{2}, \dfrac{2}{3}\right\}$

   E. $\left\{\dfrac{2}{5}, -\dfrac{2}{3}\right\}$

146   Copyright © American Book Company

Chapter 14
Graphing and Writing Equations and Inequalities

14.1 Graphing Linear Equations

In addition to graphing ordered pairs, the Cartesian plane can be used to graph the solution set for an equation. Any equation with two variables that are both to the first power is called a **linear equation.** The graph of a linear equation will always be a straight line.

Example 1: Graph the solution set for $x + y = 7$.

Step 1: Make a list of some pairs of numbers that will work in the equation.

$$\begin{array}{ll} x + y = 7 & \\ 4 + 3 = 7 & (4, 3) \\ -1 + 8 = 7 & (-1, 8) \\ 5 + 2 = 7 & (5, 2) \\ 0 + 7 = 7 & (0, 7) \end{array} \Bigg\} \text{ordered pair solutions}$$

Step 2: Plot these points on a Cartesian plane.

Step 3: By passing a line through these points, we graph the solution set for $x + y = 7$. This means that every point on the line is a solution to the equation $x + y = 7$. For example, $(1, 6)$ is a solution, so the line passes through the point $(1, 6)$.

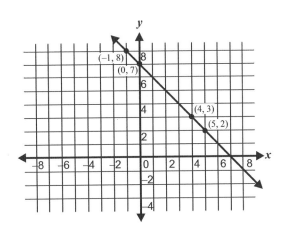

Copyright © American Book Company

Chapter 14 Graphing and Writing Equations and Inequalities

Make a table of solutions for each linear equation below. Then plot the ordered pair solutions on graph paper. Draw a line through the points. (If one of the points does not line up, you have made a mistake.)

1. $x + y = 6$
2. $y = x + 1$
3. $y = x - 2$
4. $x + 2 = y$
5. $x - 5 = y$
6. $x - y = 0$

Example 2: Graph the equation $y = 2x - 5$.

Step 1: This equation has 2 variables, both to the first power, so we know the graph will be a straight line. Substitute some numbers for x or y to find pairs of numbers that satisfy the equation. For the above equation, it will be easier to substitute values of x in order to find the corresponding value for y. Record the values for x and y in a table.

If x is 0, y would be -5
If x is 1, y would be -3
If x is 2, y would be -1
If x is 3, y would be 1

x	y
0	-5
1	-3
2	-1
3	1

Step 2: Graph the ordered pairs, and draw a line through the points.

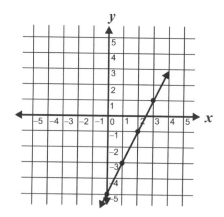

Find pairs of numbers that satisfy the equations below, and graph the line on graph paper.

1. $y = -2x + 2$
2. $2x - 2 = y$
3. $-x + 3 = y$
4. $y = x + 1$
5. $4x - 2 = y$
6. $y = 3x - 3$
7. $x = 4y - 3$
8. $2x = 3y + 1$
9. $x + 2y = 4$

14.2 Graphing Horizontal and Vertical Lines

The graph of some equations is a horizontal or a vertical line.

Example 3: $y = 3$

Step 1: Make a list of ordered pairs that satisfy the equation $y = 3$.

x	y
0	3
1	3
2	3
3	3

No matter what value of x you choose, y is always 3.

Step 2: Plot these points on an Cartesian plane, and draw a line through the points.

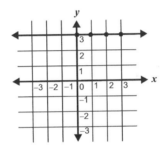

The graph is a horizontal line.

Example 4: $2x + 3 = 0$

Step 1: For these equations with only one variable, find what x equals first.
$2x + 3 = 0$
$2x = -3$
$x = \dfrac{-3}{2}$

Step 2: Using Example 3, find ordered pairs that satisfy the equation, plot the points, and graph the line.

x	y
$\dfrac{-3}{2}$	0
$\dfrac{-3}{2}$	1
$\dfrac{-3}{2}$	2
$\dfrac{-3}{2}$	3

No matter which value of y you choose, the value of x does not change.

The graph is a vertical line.

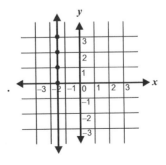

Chapter 14   Graphing and Writing Equations and Inequalities

Find pairs of numbers that satisfy the equations below, and graph the line on graph paper.

1. $2y + 2 = 0$

2. $x = -4$

3. $3x = 3$

4. $y = 5$

5. $4x - 2 = 0$

6. $2x - 6 = 0$

7. $4y = 1$

8. $5x + 10 = 0$

9. $3y + 12 = 0$

10. $x + 1 = 0$

11. $2y - 8 = 0$

12. $3x = -9$

13. $x = -2$

14. $6y - 2 = 0$

15. $5x - 5 = 0$

14.3   Finding the Distance Between Two Points

Notice that a subscript added to the x and y identifies each ordered pair uniquely in the plane. For example, point 1 is identified as (x_1, y_1), point 2 as (x_2, y_2), and so on. This unique subscript identification allows us to calculate slope, distance, and midpoints of line segments in the plane using standard formulas like the distance formula. To find the distance between two points on a Cartesian plane, use the following formula:

$$d = \sqrt{(y_2 - y_1)^2 + (x_2 - x_1)^2}$$

Example 5:   Find the distance between $(-2, 1)$ and $(3, -4)$.

Plugging the values from the ordered pairs into the formula, we find:

$$d = \sqrt{(-4 - 1)^2 + [3 - (-2)]^2}$$

$$d = \sqrt{(-5)^2 + (5)^2}$$

$$d = \sqrt{25 + 25} = \sqrt{50}$$

To simplify, we look for perfect squares that are a factor of 50. $50 = 25 \times 2$. Therefore,

$$d = \sqrt{25} \times \sqrt{2} = 5\sqrt{2}$$

Find the distance between the following pairs of points using the distance formula above.

1. $(6, -1)(5, 2)$

2. $(-4, 3)(2, -1)$

3. $(10, 2)(6, -1)$

4. $(-2, 5)(-4, 3)$

5. $(8, -2)(3, -9)$

6. $(2, -2)(8, 1)$

7. $(3, 1)(5, 5)$

8. $(-2, -1)(3, 4)$

9. $(5, -3)(-1, -5)$

10. $(6, 5)(3, -4)$

11. $(-1, 0)(-9, -8)$

12. $(-2, 0)(-6, 6)$

13. $(2, 4)(8, 10)$

14. $(-10, -5)(2, -7)$

15. $(-3, 6)(1, -1)$

14.4 Finding the Midpoint of a Line Segment

You can use the coordinates of the endpoints of a line segment to find the coordinates of the midpoint of the line segment. The formula to find the midpoint between two coordinates is:

$$\text{midpoint}, M = \left(\frac{x_1 + x_2}{2}, \frac{y_1 + y_2}{2}\right)$$

Example 6: Find the midpoint of the line segment having endpoints at $(-3, -1)$ and $(4, 3)$.
Use the formula for the midpoint. $M = \left(\frac{4 + (-3)}{2}, \frac{3 + (-1)}{2}\right)$
When we simplify each coordinate, we find the midpoint, M, is $\left(\frac{1}{2}, 1\right)$.

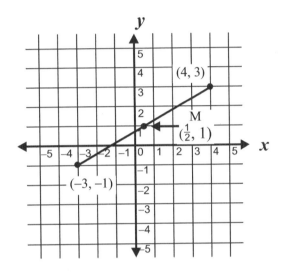

For each of the following pairs of points, find the coordinate of the midpoint, M, using the formula given above.

1. $(4, 5)\ (-6, 9)$
2. $(-3, 2)\ (-1, -2)$
3. $(3, 6)\ (9, 12)$
4. $(2, 5)\ (6, 9)$
5. $(8, 9)\ (6, 11)$

6. $(-4, 3)\ (8, 7)$
7. $(-1, -5)\ (-3, -11)$
8. $(4, 2)\ (-2, 8)$
9. $(4, 3)\ (-1, -5)$
10. $(-6, 2)\ (8, -8)$

11. $(-3, 9)\ (-9, 3)$
12. $(7, 8)\ (11, 6)$
13. $(12, 19)\ (2, 3)$
14. $(5, 4)\ (9, -2)$
15. $(-4, 6)\ (10, -2)$

Chapter 14   Graphing and Writing Equations and Inequalities

14.5   Distance Between Points

Collinear points are points that lie on the same straight line. This section teaches how to use the relationships between the collinear points to find the unknown distances between the points on the line.

To find the distance between collinear points, study the given distances to find those that are not given. Since the points are all collinear, we will only need to use a series of subtractions and/or additions for the unknown distances.

Example 7:   In the figure below, A, B, C, and D are collinear.
$\overline{AD}$ is 35 units long; $\overline{AC}$ is 22 units long, and $\overline{BD}$ is 29 units long. How many units long is $\overline{BC}$?

Method 1:   Subtract the distance $\overline{AB}$ from $\overline{AC}$.

1st step: Find the length of $\overline{AB}$: $\overline{AD} - \overline{BD} = \overline{AB}$; $35 - 29 = 6$

2nd step: Subtract $\overline{AC} - \overline{AB} = \overline{BC}$; $22 - 6 = 16$ units

Method 2:   Subtract the distance $\overline{CD}$ from $\overline{BD}$.

1st step: Find the length of $\overline{CD}$: $\overline{AD} - \overline{AC} = \overline{CD}$; $35 - 22 = 13$

2nd step: Subtract $\overline{BD} - \overline{CD} = \overline{BC}$; $29 - 13 = 16$ units

Method 3:   Subtract the sum of the distances $\overline{AB} + \overline{CD}$ from $\overline{AD}$.

1st step: use the values calculated in methods 1 and 2 to solve.

$35 - (6 + 13) = 35 - 19 = 16$ units

14.5   Distance Between Points

Read the scenario below and answer the questions that follow.

$ABCD$ is a string of aluminium wire with an unknown total length. What is known though, is that the two lengths $\overline{AB}$ and $\overline{CD}$ are equal, and measure $\sqrt{50}$ units. In an attempt to find the length of the wire, Killian reshaped it into an isosceles right angled triangle with vertices at A, B and C.

1. What point does D join with?

   A. A
   B. B
   C. C
   D. itself
   E. midpoint of $\overline{BC}$

2. What is the relationship of distance $\overline{BC}$ to distances $\overline{AB}$ and $\overline{CD}$?

   A. $\overline{BC} = \sqrt{\overline{AB} + \overline{CD}}$
   B. $2(\overline{BC}) = 2\overline{AB} - 2\overline{CD}$
   C. $(\overline{BC})^2 = (\overline{AB})^2 + (\overline{CD})^2$
   D. none
   E. $\sqrt{\overline{BC}} = \overline{AB} + \overline{CD}$

3. Which of the following is the length of $\overline{BC}$?

   A. $\sqrt[3]{50}$ units
   B. 5 units
   C. 10 units
   D. $2\sqrt{50}$ units
   E. 100 units

4. Killian has now found the length he set out to find. What is it?

   A. 75 units
   B. 20 units
   C. 250 units
   D. 110 units
   E. $10 + 2\sqrt{50}$ units

5. Find distance $\overline{BD}$?

   A. $10 + \sqrt{50}$ units
   B. 100 units
   C. 200 units
   D. 10 units
   E. $2\sqrt{50}$ units

Copyright © American Book Company

Chapter 14   Graphing and Writing Equations and Inequalities

14.6   Finding the Intercepts of a Line

The x-intercept is the point where the graph of a line crosses the x-axis. The y-intercept is the point where the graph of a line crosses the y-axis.

To find the x-intercept, set $y = 0$
To find the y-intercept, set $x = 0$

Example 8:   Find the x- and y-intercepts of the line $6x + 2y = 18$

Step 1:   To find the x-intercept, set $y = 0$.

$$6x + 2(0) = 18$$
$$\frac{6x}{6} = \frac{18}{6}$$
$$x = 3$$

The x-intercept is at the point $(3, 0)$.

Step 2:   To find the y-intercept, set $x = 0$.

$$6(0) + 2y = 18$$
$$\frac{2y}{2} = \frac{18}{2}$$
$$y = 9$$

The y-intercept is at the point $(0, 9)$.

You can now use the two intercepts to graph the line.

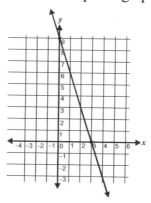

For each of the following equations, find both the x and the y intercepts of the line. For extra practice, draw each of the lines on graph paper.

1. $8x - 2y = 8$
2. $4x + 8y = 16$
3. $3x + 3y = 9$
4. $x - 2y = -5$
5. $8x + 4y = 32$

6. $3x - 4y = 12$
7. $-3x - 3y = 6$
8. $-6x + 2y = 18$
9. $4x - 2y = -4$
10. $-5x - 3y = 15$

11. $3x - 6y = -12$
12. $6x + 3y = 9$
13. $-2x - 6y = 18$
14. $2x + 3y = -6$
15. $-3x + 8y = 12$

14.7 Understanding Slope

The slope of a line refers to how steep a line is. Slope is also defined as the rate of change. When we graph a line using ordered pairs, we can easily determine the slope. Slope is often represented by the letter m.

The formula for slope of a line is: $m = \dfrac{y_2 - y_1}{x_2 - x_1}$ or $\dfrac{\text{rise}}{\text{run}}$

Example 9: What is the slope of the following line that passes through the ordered pairs $(-4, -3)$ and $(1, 3)$?

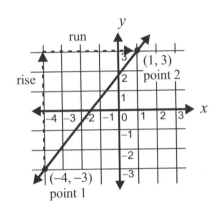

y_2 is 3, the y-coordinate of point 2.

y_1 is -3, the y-coordinate of point 1.

x_2 is 1, the x-coordinate of point 2.

x_1 is -4, the x-coordinate of point 1.

Use the formula for slope given above:

$$m = \dfrac{3 - (-3)}{1 - (-4)} = \dfrac{6}{5}$$

The slope is $\dfrac{6}{5}$. This shows us that we can go up 6 (rise) and over 5 to the right (run) to find another point on the line.

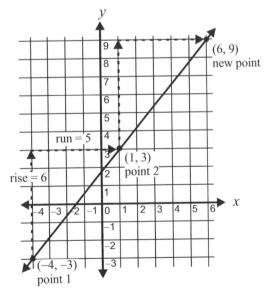

Chapter 14 Graphing and Writing Equations and Inequalities

Example 10: Find the slope of a line through the points $(-2, 3)$ and $(1, -2)$. It doesn't matter which pair we choose for point 1 and point 2. The answer is the same.

Let point 1 be $(-2, 3)$
Let point 2 be $(1, -2)$

$$\text{slope} = \frac{(y_2 - y_1)}{(x_2 - x_1)} = \frac{-2 - 3}{1 - (-2)} = \frac{-5}{3}$$

When the slope is negative, the line will slant left. For this example, the line will go **down** 5 units and then over 3 units to the **right**.

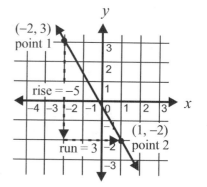

Example 11: What is the slope of a line that passes through $(1, 1)$ and $(3, 1)$?

$$\text{slope} = \frac{1 - 1}{3 - 1} = \frac{0}{2} = 0$$

When $y_2 - y_1 = 0$, the slope will equal 0, and the line will be horizontal.

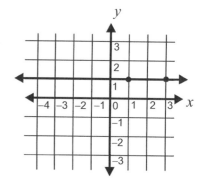

Example 12: What is the slope of a line that passes through $(2, 1)$ and $(2, -3)$?

$$\text{slope} = \frac{-3 - 1}{2 - 2} = \frac{-4}{0} = \text{undefined}$$

When $x_2 - x_1 = 0$, the slope is undefined, and the line will be vertical.

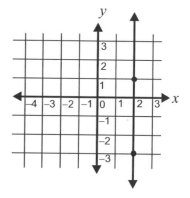

The following lines summarize what we know about slope.

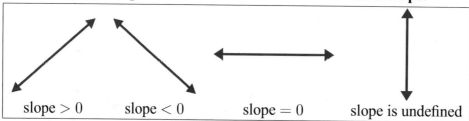

Find the slope of the line that goes through the following pairs of points. Using graph paper, graph the line through the two points, and label the rise and run. (See Examples 9–12)

1. $(2, 3)\ (4, 5)$
2. $(1, 3)\ (2, 5)$
3. $(-1, 2)\ (4, 1)$
4. $(1, -2)\ (4, -2)$
5. $(3, 0)\ (3, 4)$
6. $(3, 2)\ (-1, 8)$
7. $(4, 3)\ (2, 4)$
8. $(2, 2)\ (1, 5)$
9. $(3, 4)\ (1, 2)$
10. $(3, 2)\ (3, 6)$
11. $(6, -2)\ (3, -2)$
12. $(1, 2)\ (3, 4)$
13. $(-2, 1)\ (-4, 3)$
14. $(5, 2)\ (4, -1)$
15. $(1, -3)\ (-2, 4)$
16. $(2, -1)\ (3, 5)$

14.8 Slope-Intercept Form of a Line

An equation that contains two variables, each to the first degree, is a **linear equation**. The graph for a linear equation is a straight line. To put a linear equation in slope-intercept form, solve the equation for y. This form of the equation shows the slope and the y-intercept. Slope-intercept form follows the pattern of $y = mx + b$. The "m" represents slope, and the "b" represents the y-intercept. The y-intercept is the point at which the line crosses the y-axis.

When the slope of a line is not 0, the graph of the equation shows a **direct variation** between y and x. When y increases, x increases in a certain proportion. The proportion stays constant. The constant is called the **slope** of the line.

Example 13: Put the equation $2x + 3y = 15$ in slope-intercept form. What is the slope of the line? What is the y-intercept? Graph the line.

Step 1: Solve for y:

$$\begin{aligned} 2x + 3y &= 15 \\ -2x & -2x \end{aligned}$$

$$\frac{3y}{3} = -\frac{2x}{3} + \frac{15}{3}$$

slope-intercept form: $y = -\frac{2}{3}x + 5$

The slope is $-\frac{2}{3}$ and the y-intercept is 5.

Step 2: Knowing the slope and the y-intercept, we can graph the line.

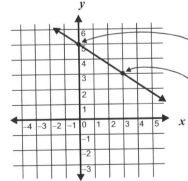

The y-intercept is 5, so the line passes through the point $(0, 5)$ on the y-axis.

The slope is $-\frac{2}{3}$, so go down 2 and over 3 to get a second point.

Chapter 14   Graphing and Writing Equations and Inequalities

Put each of the following equations in slope-intercept form by solving for y. On your graph paper, graph the line using the slope and y-intercept.

1. $4x - 5y = 5$ 6. $8x - 5y = 10$ 11. $3x - 2y = -6$ 16. $4x + 2y = 8$

2. $2x + 4y = 16$ 7. $-2x + y = 4$ 12. $3x + 4y = 2$ 17. $6x - y = 4$

3. $3x - 2y = 10$ 8. $-4x + 3y = 12$ 13. $-x = 2 + 4y$ 18. $-2x - 4y = 8$

4. $x + 3y = -12$ 9. $-6x + 2y = 12$ 14. $2x = 4y - 2$ 19. $5x + 4y = 16$

5. $6x + 2y = 0$ 10. $x - 5y = 5$ 15. $6x - 3y = 9$ 20. $6 = 2y - 3x$

14.9   Verify That a Point Lies on a Line

To know whether or not a point lies on a line, substitute the coordinates of the point into the formula for the line. If the point lies on the line, the equation will be true. If the point does not lie on the line, the equation will be false.

Example 14:   Does the point $(5, 2)$ lie on the line given by the equation $x + y = 7$?

Solution: Substitute 5 for x and 2 for y in the equation. $5 + 2 = 7$. Since this is a true statement, the point $(5, 2)$ does lie on the line $x + y = 7$.

Example 15:   Does the point $(0, 1)$ lie on the line given by the equation $5x + 4y = 16$?

Solution: Substitute 0 for x and 1 for y in the equation $5x + 4y = 16$. Does $5(0) + 4(1) = 16$? No, it equals 4, not 16. Therefore, the point $(0, 1)$ is not on the line given by the equation $5x + 4y = 16$.

For each point below, state whether or not it lies on the line given by the equation that follows the point coordinates.

1. $(2, 4)$ $6x - y = 8$ 5. $(3, 7)$ $x - 5y = -32$ 9. $(6, 8)$ $6x - y = 28$

2. $(1, 1)$ $6x - y = 5$ 6. $(0, 5)$ $-6x - 5y = 3$ 10. $(-2, 3)$ $x + 2y = 4$

3. $(3, 8)$ $-2x + y = 2$ 7. $(2, 4)$ $4x + 2y = 16$ 11. $(4, -1)$ $-x - 3y = -1$

4. $(9, 6)$ $-2x + y = 0$ 8. $(9, 1)$ $3x - 2y = 29$ 12. $(-1, -3)$ $2x + y = 1$

158   Copyright © American Book Company

14.10 Graphing a Line Knowing a Point and Slope

If you are given a point of a line and the slope of a line, the line can be graphed.

Example 16: Given that line l has a slope of $\frac{4}{3}$ and contains the point $(2, -1)$, graph the line.

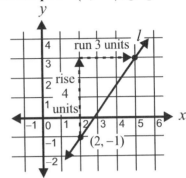

Plot and label the point $(2, -1)$ on a Cartesian plane.

The slope, m, is $\frac{4}{3}$, so the rise is 4, and the run is 3. From the point $(2, -1)$, count 4 units up and 3 units to the right.

Draw the line through the two points.

Example 17: Given a line that has a slope of $-\frac{1}{4}$ and passes through the point $(-3, 2)$, graph the line.

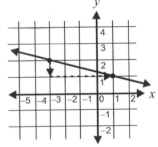

Plot the point $(-3, 2)$.

Since the slope is negative, go **down** 1 unit and over 4 units to get a second point.

Graph the line through the two points.

Graph a line on your own graph paper for each of the following problems. First, plot the point. Then use the slope to find a second point. Draw the line formed from the point and the slope.

1. $(2, -2)$, $m = \frac{3}{4}$
2. $(3, -4)$, $m = \frac{1}{2}$
3. $(1, 3)$, $m = -\frac{1}{3}$
4. $(2, -4)$, $m = 1$
5. $(3, 0)$, $m = -\frac{1}{2}$
6. $(-2, 1)$, $m = \frac{4}{3}$
7. $(-4, -2)$, $m = \frac{1}{2}$
8. $(1, -4)$, $m = \frac{3}{4}$
9. $(2, -1)$, $m = -\frac{1}{2}$
10. $(5, -2)$, $m = \frac{1}{4}$
11. $(-2, -3)$, $m = \frac{2}{3}$
12. $(4, -1)$, $m = -\frac{1}{3}$
13. $(-1, 5)$, $m = \frac{2}{5}$
14. $(-2, 3)$, $m = \frac{3}{4}$
15. $(4, 4)$, $m = -\frac{1}{2}$
16. $(3, -3)$, $m = -\frac{3}{4}$
17. $(-2, 5)$, $m = \frac{1}{3}$
18. $(-2, -3)$, $m = -\frac{3}{4}$
19. $(4, -3)$, $m = \frac{2}{3}$
20. $(1, 4)$, $m = -\frac{1}{2}$

Chapter 14   Graphing and Writing Equations and Inequalities

14.11   Finding the Equation of a Line Using the Slope and Y-Intercept

Find the equation of a line is really simple if you are given two things: the slope and the y-intercept. First you need to know that the easiest equation to find when given these two things is the equation in slope-intercept form. Slope intercept form is $y = mx + b$, where m is the slope and b is the y-intercept.

Example 18:   Find the equation of a line given $m = \frac{1}{2}$ and y-intercept $= 4$.

$$y = mx + b \quad \longrightarrow \quad \text{Slope-Intercept form}$$

$$y = \tfrac{1}{2}x + 4 \quad \longrightarrow \quad \text{Plug in values for } m \text{ and } b.$$

Example 19:   Find the equation of a line that has a slope of 3 and a y-intercept of $-\frac{1}{3}$.

$$y = mx + b$$

$$y = 3x - \tfrac{1}{3}$$

Find the equation using the slope and y-intercept.

1. $m = 2$, y-intercept $= 4$

2. $m = 6$, y-intercept $= -2$

3. $m = -1$, y-intercept $= 7$

4. $m = -3$, y-intercept $= 0$

5. $m = -\frac{1}{2}$, y-intercept $= -11$

6. $m = \frac{2}{3}$, y-intercept $= 13$

7. $m = \frac{5}{4}$, y-intercept $= -\frac{4}{5}$

8. $m = -\frac{9}{2}$, y-intercept $= \frac{3}{9}$

9. $m = 1$, y-intercept $= -\frac{7}{5}$

10. $m = 0$, y-intercept $= 2$

11. $m = 2$, y-intercept $= -4$

12. $m = -7$, y-intercept $= 11$

13. $m = \frac{1}{2}$, y-intercept $= -5$

14. $m = 5$, y-intercept $= 6$

15. $m = -\frac{4}{5}$, y-intercept $= 8$

16. $m = 9$, y-intercept $= 1$

17. $m = \frac{3}{4}$, y-intercept $= 0$

18. $m = 7$, y-intercept $= -\frac{1}{4}$

19. $m = \frac{1}{7}$, y-intercept $= 12$

20. $m = \frac{6}{7}$, y-intercept $= \frac{7}{9}$

160   Copyright © American Book Company

14.12 Finding the Equation of a Line Using Two Points or a Point and Slope

14.12 Finding the Equation of a Line Using Two Points or a Point and Slope

If you can find the slope of a line and know the coordinates of one point, you can write the equation for the line. You know the formula for the slope of a line is:

$$m = \frac{y_2 - y_1}{x_2 - x_1} \text{ or } \frac{y_2 - y_1}{x_2 - x_1} = m$$

Using algebra, you can see that if you multiply both sides of the equation by $x_2 - x_1$, you get:

$$y - y_1 = m(x - x_1) \longleftarrow \text{point-slope form of an equation}$$

Example 20: Write the equation of the line passing through the points $(-2, 3)$ and $(1, 5)$.

Step 1: First, find the slope of the line using the two points given.

$$m = \frac{y_2 - y_1}{x_2 - x_1} = \frac{5 - 3}{1 - (-2)} = \frac{2}{3}$$

Step 2: Pick one of the two points to use in the point-slope equation. For point $(-2, 3)$, we know $x_1 = -2$ and $y_1 = 3$, and we know $m = \frac{2}{3}$. Substitute these values into the point-slope form of the equation.

$$y - y_1 = m(x - x_1)$$

$$y - 3 = \frac{2}{3}[x - (-2)]$$

$$y - 3 = \frac{2}{3}x + \frac{4}{3}$$

$$y = \frac{2}{3}x + \frac{13}{3}$$

Use the point-slope formula to write an equation for each of the following lines.

1. $(1, -2), m = 2$

2. $(-3, 3), m = \frac{1}{3}$

3. $(4, 2), m = \frac{1}{4}$

4. $(5, 0), m = 1$

5. $(3, -4), m = \frac{1}{2}$

6. $(-1, -4) \ (2, -1)$

7. $(2, 1) \ (-1, -3)$

8. $(-2, 5) \ (-4, 3)$

9. $(-4, 3) \ (2, -1)$

10. $(3, 1) \ (5, 5)$

11. $(-3, 1), m = 2$

12. $(-1, 2), m = \frac{4}{3}$

13. $(2, -5), m = -2$

14. $(-1, 3), m = \frac{1}{3}$

15. $(0, -2), m = -\frac{3}{2}$

Copyright © American Book Company

Chapter 14 Graphing and Writing Equations and Inequalities

14.13   Matching Graphs of Linear Equations

Match each equation below with the graph of the equation.

A. $x = -4$
B. $x = y$
C. $-\frac{1}{2}x = y$
D. $y = -4$
E. $4x + y = 4$
F. $y = x - 3$
G. $x - 2y = 6$
H. $2x + 3y = 6$
I. $y = 3x + 2$

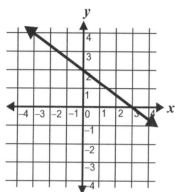

1. _____

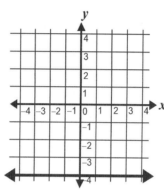

4. _____

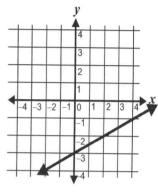

7. _____

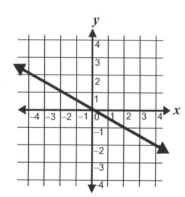

2. _____

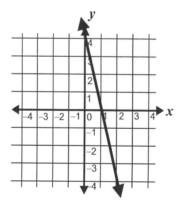

5. _____

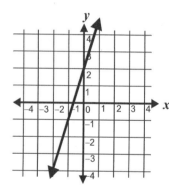

8. _____

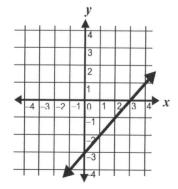

3. _____

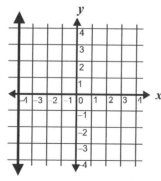

6. _____

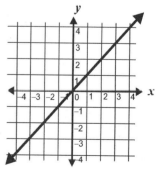

9. _____

162   Copyright © American Book Company

14.14 Graphing Inequalities

In the previous section, you would graph the equation $x = 3$ as:

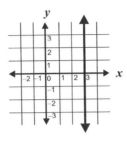

In this section, we graph inequalities such as $x > 3$ (read x is greater than 3). To show this, we use a broken line since the points on the line $x = 3$ are not included in the solution. We shade all points greater than 3.

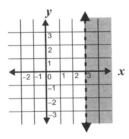

When we graph $x \geq 3$ (read x is greater than or equal to 3), we use a solid line because the points on the line $x = 3$ are included in the graph.

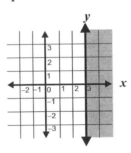

Graph the following inequalities on your own graph paper.

1. $y < 2$
2. $x \geq 4$
3. $y \geq 1$
4. $x < -1$
5. $y \geq -2$
6. $x \leq -4$
7. $x > -3$
8. $y \leq 3$
9. $x \leq 5$
10. $y > -5$
11. $x \geq 3$
12. $y < -1$
13. $x \leq 0$
14. $y > -1$
15. $y \leq 4$
16. $x \geq 0$
17. $y \geq 3$
18. $x < 4$
19. $x \leq -2$
20. $y < -2$
21. $y \geq -4$
22. $x \geq -1$
23. $y \leq 5$
24. $x < -3$

Chapter 14 Graphing and Writing Equations and Inequalities

Example 21: Graph $x + y \geq 3$.

Step 1: First, we graph $x + y \geq 3$ by changing the inequality to an equality. Think of ordered pairs that will satisfy the equation $x + y = 3$. Then, plot the points, and draw the line. As shown below, this line divides the Cartesian plane into 2 half-planes, $x + y \geq 3$ and $x + y \leq 3$. One half-plane is above the line, and the other is below the line.

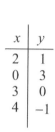
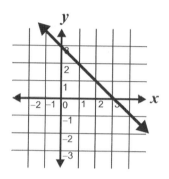

Step 2: To determine which side of the line to shade, first choose a test point. If the point you choose makes the inequality true, then the point is on the side you shade. If the point you choose does not make the inequality true, then shade the side that does not contain the test point.

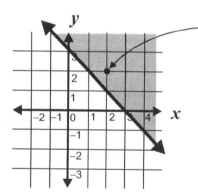

For our test point, let's choose $(2, 2)$. Substitute $(2, 2)$ into the inequality.

$x + y \geq 3$
$2 + 2 \geq 3$

$4 \geq 3$ is true, so shade the side that includes this point.

Use a solid line because of the $\geq$ sign.

Graph the following inequalities on your own graph paper.

1. $x + y \leq 4$
2. $x + y \geq 3$
3. $x \geq 5 - y$
4. $x \leq 1 + y$
5. $x - y \geq -2$
6. $x < y + 4$
7. $x + y < -1$
8. $x - y \leq 0$
9. $x \geq y + 2$
10. $x < -y + 1$
11. $-x + y > 1$
12. $-x - y < -2$

14.14 Graphing Inequalities

For more complex inequalities, it may be easier to graph by first changing the inequality to an equality and then put the equation in slope-intercept form.

Example 22: Graph the inequality $2x + 4y \leq 8$.

Step 1: Change the inequality to an equality.

$2x + 4y = 8$

Step 2: Put the equation in slope-intercept form by solving the equation for y.

$$2x + 4y = 8$$
$$2x - 2x + 4y = -2x + 8 \quad \text{Subtract } 2x \text{ from both sides of the equation.}$$
$$4y = -2x + 8 \quad \text{Simplify.}$$
$$\frac{4y}{4} = \frac{-2x + 8}{4} \quad \text{Divide both sides by 4.}$$
$$y = \frac{-2x}{4} + \frac{8}{4} \quad \text{Find the lowest terms of the fractions.}$$
$$y = -\tfrac{1}{2}x + 2$$

Step 3: Graph the line. If the inequality is $<$ or $>$, use a dotted line. If the inequality is $\leq$ or $\geq$, use a solid line. For this example, we should use a solid line.

Step 4: Determine which side of the line to shade. Pick a point such as $(0,0)$ to see if it is true in the inequality.

$2x + 4y \leq 8$, so substitute $(0,0)$.
Is $0 + 0 \leq 8$? Yes, $0 \leq 8$, so shade the side of the line that includes the point $(0,0)$.

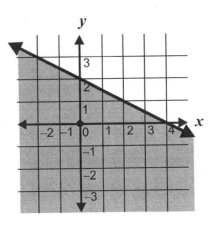

Graph the following inequalities on your own graph paper.

1. $2x + y \geq 1$
2. $3x - y \leq 3$
3. $x + 3y > 12$
4. $4x - 3y < 12$
5. $y \geq 3x + 1$
6. $x - 2y > -2$
7. $x \leq y + 4$
8. $x + y < -1$
9. $-4y \geq 2x + 1$
10. $x \leq 4y - 2$
11. $3x - y \geq 4$
12. $y \geq 2x - 5$

Chapter 14 Review

1. Graph the solution set for the linear equation: $x - 3 = y$.

2. Graph the equation $2x - 4 = 0$.

3. What is the slope of the line that passes through the points $(5, 3)$ and $(6, 1)$?

4. What is the slope of the line that passes through the points $(-1, 4)$ and $(-6, -2)$?

5. What is the x-intercept for the following equation? $6x - y = 30$

6. What is the y-intercept for the following equation? $4x + 2y = 28$

7. Graph the equation $3y = 9$.

8. Write the following equation in slope-intercept form.
$3x = -2y + 4$

9. What is the slope of the line $y = -\frac{1}{2}x + 3$?

10. What is the x-intercept of the line $y = 5x + 6$?

11. What is the y-intercept of the line $y - \frac{2}{3}x + 3 = 0$?

12. Graph the line which has a slope of -2 and a y-intercept of -3.

13. Find the equation of the line which contains the point $(0, 2)$ and has a slope of $\frac{3}{4}$.

14. What is the distance between the points $(3, 3)$ and $(6, -1)$?

15. What is the distance between the two points $(-3, 0)$ and $(2, 5)$?

16. Find the midpoint of the two points $(6, 10)$ and $(-4, 4)$.

17. Find the midpoint of the two points $(-1, -7)$ and $(5, 3)$.

Find the equation using the slope and y-intercept.

18. $m = 2$, y-intercept $= \frac{1}{4}$

19. $m = -4$, y-intercept $= -12$

20. $m = \frac{1}{2}$, y-intercept $= 17$

21. $m = -\frac{5}{2}$, y-intercept $= \frac{3}{2}$

Graph the following inequalities on a Cartesian plane using graph paper.

22. $x \geq 4$

23. $x \leq -2$

24. $5y > -10x + 5$

25. $y - 2x \leq 3$

Find the distance between the points.

26. The points A, B, C, and D are collinear points; C is the midpoint of $\overline{AD}$, while B is the midpoint of $\overline{AC}$. Given that $\overline{AB}$ is 7 units, find the length of $\overline{CD}$?

27. Neighbors Adams, Carters, Dawkins, Edwards, and Beckhams live on one side of a straight road measuring 1 mile long. Each home has it's mail box by the roadside, and next to the driveway. The Beckhams live at one end of the road and their mailbox is 3,000 feet from the Dawkins. The driveways of the Carters and Dawkins homes are 720 feet apart, while the driveways between the Edwards and Adams is 1,000 feet.

1 mile = 5280 feet

How far apart is the Adams' mailbox from the Carter's mailbox?

Chapter 14 Test

1. Which is the graph of $x - 3y = 6$?

   A.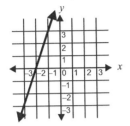

   B.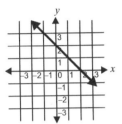

   C.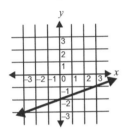

   D.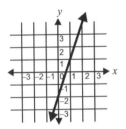

   E.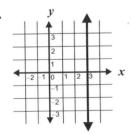

2. Which of the following points does not lie on the line $y = 3x - 2$?
   - F. $(0, -2)$
   - G. $(1, 1)$
   - H. $(1, 5)$
   - J. $(2, 4)$
   - K. $(-1, -5)$

3. Which of the following is the graph of the equation $y = x - 3$?

   A.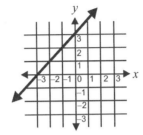

   B.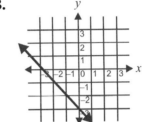

   C.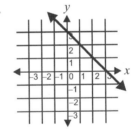

   D.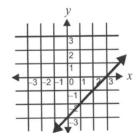

   E.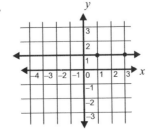

Chapter 14 Graphing and Writing Equations and Inequalities

4. What is the x-intercept of the following linear equation? $3x + 4y = 12$

F. $(0, 3)$
G. $(3, 0)$
H. $(0, 4)$
J. $(4, 0)$
K. $(3, 4)$

5. Which of the following equations is represented by the graph?

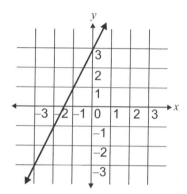

A. $y = -3x + 3$
B. $y = -\frac{1}{3}x + 3$
C. $y = 3x - 3$
D. $y = 2x + 3$
E. $y = x + 3$

6. What is the equation of the line that includes the point $(4, -3)$ and has a slope of -2?

F. $y = -2x - 5$
G. $y = -2x - 2$
H. $y = -2x + 5$
J. $y = 2x - 5$
K. $y = 2x + 5$

7. What is the x-intercept and y-intercept for the equation $x + 2y = 6$?

A. x-intercept $= (0, 6)$
   y-intercept $= (3, 0)$
B. x-intercept $= (4, 1)$
   y-intercept $= (2, 2)$
C. x-intercept $= (0, 6)$
   y-intercept $= (0, 3)$
D. x-intercept $= (6, 0)$
   y-intercept $= (0, 3)$
E. x-intercept $= (6, 0)$
   y-intercept $= (3, 0)$

8. Look at the graphs below. Which of the following statements is false?

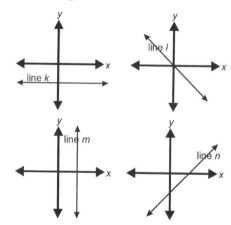

F. The slope of line k is undefined.
G. The slope of line l is negative.
H. The slope of line m is undefined.
J. The slope of line n is positive.
K. All of the statements above are true.

9. Which equation is represented by $m = -2$ and y-intercept $= 2$?

A. $y = -2x - 2$
B. $y = -2x + 2$
C. $y = 2x + 2$
D. $y = 2x - 2$
E. $y = 2x + 1$

Chapter 14 Test

10. Which of the following graphs shows a line with a slope of 0 that passes through the point $(3, 2)$?

F.
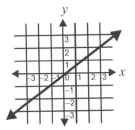

G.
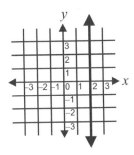

H.
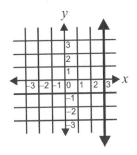

J.
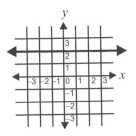

K.
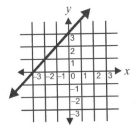

11. Put the following equation in slope-intercept form: $2x - 3y = 6$

A. $y = -\frac{2}{3}x - 2$

B. $y = 2x - 2$

C. $y = -\frac{2}{3}x + 2$

D. $y = 2x + 2$

E. $y = \frac{2}{3}x - 2$

12. What is the y-intercept of $y = 2x - 6$?

F. 6
G. -6
H. 2
J. -2
K. $\frac{1}{3}$

13. Which equation passes through the points $(-1, 6)$ and $(0, 2)$?

A. $y = 4x + 2$
B. $y = -4x - 2$
C. $y = 2x + 4$
D. $y = -2x + 4$
E. $y = -4x + 2$

14. Which of the following is a solution of $3x = 5y + 1$?

F. $(3, 2)$
G. $(8, 5)$
H. $\left(-\frac{1}{3}, 0\right)$
J. $(-2, -1)$
K. $(7, 4)$

15. $(-2, 1)$ is a solution for which of the following equations?

A. $y + 2x = 4$
B. $-2x - y = 5$
C. $x + 2y = -4$
D. $2x - y = -5$
E. $3x + 1 = 7$

Copyright © American Book Company

Chapter 14   Graphing and Writing Equations and Inequalities

16. Which of the following is a graph of the inequality $-y \geq 2$?

F.
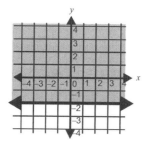

G.
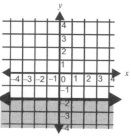

H.
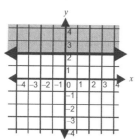

J.
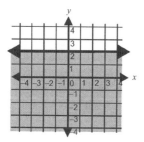

K.
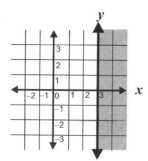

17. Which of the following is a graph of the inequality $y \leq x - 3$?

A.
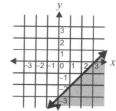

B.
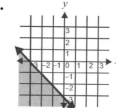

C.

D.
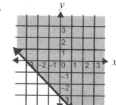

E.
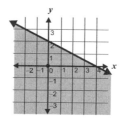

Chapter 14 Test

Read the scenario below and answer the questions that follow.

Part of the U.S. and Canadian border is a 1,200 mile dead straight line which runs from a city called Pembina in the east, to one called Blaine in the west. Other cities, too are dotted between these two cities along this border. The city of Oroville is 151 miles east of Blaine, and Midway is 46 miles from Oroville going east. While 67 miles east of Midway is the beautiful little city of Nelway.

The city of Walhalla is 38 miles from Pembina and 183 miles from Walhalla, going further west is Sherwood. To the city of Crosby from Sherwood, one has to travel 74 miles in the westerly direction.

18. In what direction is Nelway from Crosby?

   F. West

   G. East

   H. North-East

   J. North

   K. South

19. Find the distance between Oroville and Pembina.

   A. 1,351 miles

   B. 556 miles

   C. 93 miles

   D. 1,049 miles

   E. 66 miles

20. How far from Walhalla is the little beautiful city of Nelway?

   F. 887 miles

   G. 898 miles

   H. not enough information

   J. 264 miles

   K. 38 miles

21. Which of the following cities is furthest from Blaine?

   A. Midway

   B. Oroville

   C. Sherwood

   D. Walhalla

   E. Crosby

22. Of the pairs of cities given below, which pair has the shortest distance between them?

   F. Blaine - Walhalla

   G. Oroville - Sherwood

   H. Nelway - Crosby

   J. Oroville - Pembina

   K. Midway - Crosby

Copyright © American Book Company

171

Chapter 15
Systems of Equations

15.1 Equations of Parallel Lines

If two linear equations have the same slope but different y-intercepts, they are **parallel** lines. Parallel lines never touch each other, so they have no points in common.

Example 1: Consider line l shown in Figure 2 at right. The equation of the line is $y = -\frac{1}{2}x + 3$. What happens to the graph of the line if the y-intercept is changed to -1?

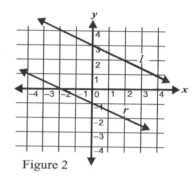

Figure 2

Rewrite the equation of the line replacing the y-intercept with -1. The equation of the new line is $y = -\frac{1}{2}x - 1$.

Graph the new line. Line r in Figure 2 is the graph of the equation $y = -\frac{1}{2}x - 1$. Since both lines l and r have the same slope, they are parallel. Line r, with a y-intercept of -1, sits below line l, with a y-intercept of 3.

Put each pair of the following equations in slope-intercept form. Write P if the lines are parallel and NP if the lines are not parallel.

1. $y = x + 1$ _____
   $2y - x = 6$

2. $3x + y = 4$ _____
   $3x = 7 - y$

3. $x + 6y = 0$ _____
   $6y + 6 = x$

4. $y = 3 - \frac{1}{2}x$ _____
   $2y + x = -6$

5. $x = 5y$ _____
   $-x = -5y + 14$

6. $y = x + 11$ _____
   $-y = x + 9$

7. $y = 3 - \frac{1}{3}x$ _____
   $2x + 3y = 3$

8. $x + y = 7$ _____
   $7 - y = 2x$

9. $x - 8y = 0$ _____
   $8y = x - 8$

10. Anton graphs the function $y = 2x - 1$. Is the graph of the function $y = 2x + 5$ parallel to Anton's graph?

11. Havana graphs the function $2y = x - 9$. Is the graph of the function $y - 2x + 2$ parallel to Havana's graph?

12. Jacques graphs the function $y - x = 4$. Is the graph of the function $y = x - 1$ parallel to Jacques's graph?

15.2 Equations of Perpendicular Lines

Now that we know how to calculate the slope of lines using two points, we are going to learn how to calculate the slope of a line perpendicular to a given line, then find the equation of that perpendicular line. To find the slope of a line perpendicular to any given line, take the slope of the first line, m:

1. multiply the slope by -1
2. invert (or flip over) the slope

You now have the slope of a perpendicular line. Writing the equation for a line perpendicular to another line involves three steps:

1. Find the slope of the perpendicular line.
2. Choose one point on the first line.
3. Use the point-slope form to write the equation.

Example 2: The solid line on the graph below has a slope of $\frac{2}{3}$. Write the equation of a line perpendicular to the solid line.

Step 1: Find the slope of the solid line. Multiply the slope by -1 and then find the inverse (flip it over).

$$\frac{2}{3} \times -1 = -\frac{2}{3} \curvearrowright -\frac{3}{2}$$

The slope of the perpendicular line, shown as a dotted line on the graph below, is $-\frac{3}{2}$.

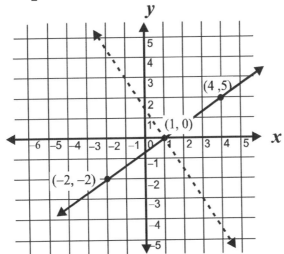

Step 2: Choose one point on the first line. We will use $(1, 0)$ in this example. The point $(-2, -2)$ or $(4, 5)$ could also be used.

Step 3: Use the point-slope formula, $(y - y_1) = m(x - x_1)$, to write the equation of the perpendicular line. Remember, we chose $(1, 0)$ as our point. So, $(y - 0) = -\frac{3}{2}(x - 1)$. Simplified, $y = -\frac{3}{2}x + \frac{3}{2}$.

Chapter 15   Systems of Equations

Solve the following problems involving perpendicular lines.

1. Find the slope of the line perpendicular to the line shown below, and draw the perpendicular as a dotted line. Use the point $(-1, 0)$ on the line and the calculated slope to find the equation of the perpendicular line.

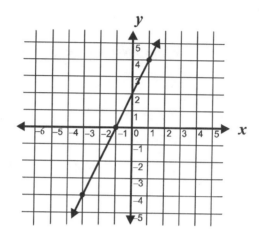

Find the equation of a line that is perpendicular to the given line using the point and slope given and the formula $(y - y_1) = m(x - x_1)$.

2. $(2, 1), 5$

3. $(3, 2), 2$

4. $(-2, 1), -3$

5. $(-4, 2), -\dfrac{1}{2}$

6. $(-1, 4), 1$

7. $(3, 3), \dfrac{2}{3}$

8. $(5, -1), -1$

9. $\left(\dfrac{1}{2}, \dfrac{3}{4}\right), 4$

10. $\left(\dfrac{2}{3}, \dfrac{3}{4}\right), -\dfrac{1}{6}$

11. $(7, -2), -\dfrac{1}{8}$

12. $(5, 0), \dfrac{4}{5}$

13. $(-3, -3), -\dfrac{7}{3}$

14. $\left(\dfrac{1}{4}, 4\right), \dfrac{1}{2}$

15. $(0, 6), -\dfrac{1}{9}$

16. Arthur graphs the function $y = 3x - 1$. Is the graph of the function $y = -3x + 5$ perpendicular to Arthur's graph?

17. Mieko graphs the function $y = x + 9$. Is the graph of the function $y = -x + 9$ perpendicular to Mieko's graph?

18. Tamar graphs the function $y = \tfrac{1}{5}x - 2$. Is the graph of the function $y = -5x + 1$ perpendicular to Tamar's graph?

15.3 Systems of Equations

Two linear equations considered at the same time are called a **system** of linear equations. The graph of a linear equation is a straight line. The graphs of two linear equations can show that the lines are **parallel**, **intersecting**, or **identical**. Two lines that are **parallel** will never intersect and have no ordered pairs in common. If two lines are **intersecting**, they have one point in common, and in this chapter, you will learn to find the ordered pair for that point. If the graph of two linear equations is the same line, the lines are said to be **identical**.

If you are given a system of two linear equations, and you put both equations in slope-intercept form, you can immediately tell if the graph of the lines will be **parallel**, **intersecting**, or **identical**.

If two linear equations have the same slope and the same y-intercept, then they are both equations for the same line. They are called **identical** or **collinear** lines. A line is made up of an infinite number of points extending infinitely far in both directions. Therefore, identical lines have an infinite number of points in common.

Example 3: $2x + 3y = -3$ **In slope intercept form:** $y = -\frac{2}{3}x - 1$

$4x + 6y = -6$ **In slope intercept form:** $y = -\frac{2}{3}x - 1$

The slope and y-intercept of both lines are the same. They are identical.

If two linear equations have the same slope but different y-intercepts, they are **parallel** lines. Parallel lines never touch each other, so they have no points in common.

If two linear equations have different slopes, then they are intersecting lines and share exactly one point in common.

The chart below summarizes what we know about the graphs of two equations in slope-intercept form.

y-Intercepts	Slopes	Graphs	Number of Solutions
same	same	identical	infinite
different	same	distinct parallel lines	none (they never touch)
same or different	different	intersecting lines	exactly one

Copyright © American Book Company

Chapter 15   Systems of Equations

For the pairs of equations below, put each equation in slope-intercept form, and tell whether the graphs of the lines will be identical, parallel, or intersecting. Also, determine how many solutions the system of equations will have.

1. $3y = 2x + 9$
   $18 = 6y - 4x$

2. $-x + y = -5$
   $x - y = 5$

3. $y = 3x + 2$
   $y - 3x = 2$

4. $-x = y$
   $-x = 2 + y$

5. $x + y = 4$
   $-x + y = 4$

6. $3x = y + 1$
   $y = 3x + 1$

7. $2x - y = 4$
   $-4x + 2y = -8$

8. $3x + y = 1$
   $x + y = 1$

9. $-y = x - 7$
   $y + x = -7$

10. $10x - 5y = 3$
    $5x - 10y = 3$

11. $-2x + 3y = 5$
    $x = 2 - y$

12. $4x - 3y = 12$
    $y = \frac{4}{3}x - 4$

13. $2x + 2y = 18$
    $y + x = 9$

14. $3x - 7y = 10$
    $6x - 14y = 20$

15. $2x = 4y - 1$
    $7y = x - 7$

16. $8y = x - 5$
    $y - \frac{1}{8}x = 12$

17. $3x - y = 1$
    $2y = -6x + 5$

18. $9 = 3x - y$
    $x = y + 3$

19. $-2x = y - 5$
    $x - 5 = 2y$

20. $\frac{1}{2}x - y$
    $-y = -\frac{1}{2}x$

176   Copyright © American Book Company

15.4 Finding Common Solutions for Intersecting Lines

When two lines intersect, they have exactly one point in common.

Example 4: $4x + y = 3$ and $x - y = 1$

Put each equation in slope-intercept form.

$$4x + y = 3 \qquad\qquad x - y = 1$$
$$y = -4x + 3 \qquad\qquad -y = -x + 1$$
$$\qquad\qquad\qquad\qquad\qquad y = x - 1$$

slope-intercept form

Straight lines with different slopes are **intersecting lines**. Look at the graphs of the lines on the same Cartesian plane.

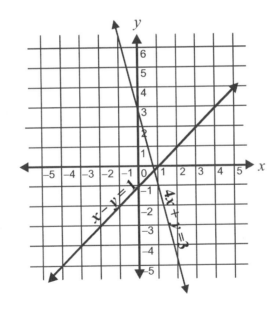

You can see from looking at the graph that the intersecting lines have one point in common. However, it is hard to tell from looking at the graph what the coordinates are for the point of intersection. To find the exact point of intersection, you can use the **substitution method** to solve the system of equations algebraically.

Copyright © American Book Company

177

Chapter 15   Systems of Equations

15.5   Solving Systems of Equations by Substitution

You can solve systems of equations by using the substitution method.

Example 5:   Find the point of intersection of the following two equations:

Equation 1: $x - y = 1$

Equation 2: $x + 3y = 9$

Step 1:   Solve one of the equations for x or y. Let's choose to solve equation 1 for x.

Equation 1: $x - y = 1$

$x = y + 1$

Step 2:   Substitute the value of x from equation 1 in place of x in equation 2.

Equation 2: $x + 3y = 9$

$(y + 1) + 3y = 9$

$4y + 1 = 9$

$4y = 9 - 1$

$4y = 8$

$y = 2$

Step 3:   Substitute the solution for y back into either equation and solve for x. We are using equation 1.

Equation 1: $x - y = 1$

$x - 2 = 1$

$x = 3$

Step 4:   The solution set is $(3, 2)$. Substitute the point into both of the equations to check.

Equation 1: $x - y = 1$ Equation 2: $x + 3y = 9$

$3 - 2 = 1$ $3 + 3(2) = 9$

$1 = 1$ $3 + 6 = 9$

$9 = 9$

The point $(3, 2)$ is common for both equations. This is the **point of intersection**.

178   Copyright © American Book Company

15.5   Solving Systems of Equations by Substitution

For each of the following pairs of equations, find the point of intersection, the common solution, using the substitution method.

1. $x + y = 22$
   $x - y = 8$

2. $y - 5 = x$
   $y - 7 = 2x$

3. $y = 2 - x$
   $5 = y + 2x$

4. $y - 9 = -x$
   $-7 = y - 3x$

5. $y = x + 2$
   $y + 1 = 2x$

6. $5y - 8 = x$
   $3y - 4 = x$

7. $2y + 6 = x$
   $y - 3x = -13$

8. $y + 3 = 2x$
   $y + 5 = 3x$

9. $y - \frac{4}{5}x = 1$
   $y - x = 1$

10. $y + \frac{1}{3}x = -1$
    $y + 1 = 2x$

11. $y - 5 = 2x$
    $3y + x = 8$

12. $x - y = 0$
    $y + 1 = 2x$

13. $x - y = 1$
    $\frac{1}{2}x + y = 5$

14. $y + x = 1$
    $y + 4x = 10$

15. $y + x = -1$
    $y - 2x = -16$

Copyright © American Book Company

179

Chapter 15   Systems of Equations

15.6   Solving Systems of Equations by Adding or Subtracting

You can solve systems of equations algebraically by adding or subtracting an equation from another equation or system of equations.

Example 6:   Find the point of intersection of the following two equations:
Equation 1: $x + y = 10$
Equation 2: $-x + 4y = 5$

Step 1:   Eliminate one of the variables by adding the two equations together. Since the x has the same coefficient in each equation, but opposite signs, it will cancel nicely by adding.

$$\begin{array}{r} x + y = 10 \\ + (-x + 4y = 5) \\ \hline 0 + 5y = 15 \\ 5y = 15 \\ y = 3 \end{array}$$

Add each like term together.
Simplify.
Divide both sides by 5.

Step 2:   Substitute the solution for y back into an equation, and solve for x.

$$\begin{array}{ll} \text{Equation 1:} & x + y = 10 \quad \text{Substitute 3 for } y. \\ & x + 3 = 10 \quad \text{Subtract 3 from both sides.} \\ & x = 7 \end{array}$$

Step 3:   The solution set is $(7, 3)$. To check, substitute the solution into both of the original equations.

Equation 1:	$x + y = 10$	Equation 2:	$-x + 4y = 5$
	$7 + 3 = 10$		$-(7) + 4(3) = 5$
	$10 = 10$		$-7 + 12 = 5$
			$5 = 5$

The point $(7, 3)$ is the point of intersection.

Example 7:   Find the point of intersection of the following two equations:
Equation 1: $3x - 2y = -1$
Equation 2: $-4y = -x - 7$

Step 1:   Put the variables in equation 2 on the same side.

$$\begin{array}{ll} -4y = -x - 7 & \text{Add } x \text{ to both sides.} \\ x - 4y = -x + x - 7 & \text{Simplify.} \\ x - 4y = -7 & \end{array}$$

Step 2:   Add the two equations together to cancel one variable. Since each variable has the same sign and different coefficients, we have to multiply one equation by a negative number so one of the variables will cancel. Equation 1's y variable has a coefficient of 2, and if multiplied by -2, the y will have the same variable as the y in equation 2, but a different sign. This will cancel nicely when added.

$$\begin{array}{ll} -2(3x - 2y = -1) & \text{Multiply by } -2. \\ -6x + 4y = 2 & \end{array}$$

180   Copyright © American Book Company

15.6   Solving Systems of Equations by Adding or Subtracting

Step 3: Add the two equations.

$$-6x + 4y = 2$$
$$\underline{+ (x \quad 4y = -7)} \qquad \text{Add equation 2 to equation 1.}$$
$$-5x + 0 = -5 \qquad \text{Simplify.}$$
$$-5x = -5 \qquad \text{Divide both sides by } -5.$$
$$x = 1$$

Step 4: Substitute the solution for x back into an equation and solve for y.

Equation 1: $\qquad 3x - 2y = -1 \qquad$ Substitute 1 for x.
$$3(1) - 2y = -1 \qquad \text{Simplify.}$$
$$3 - 2y = -1 \qquad \text{Subtract 3 from both sides.}$$
$$3 - 3 - 2y = -1 - 3 \qquad \text{Simplify.}$$
$$-2y = -4 \qquad \text{Divide both sides by } -2.$$
$$y = 2$$

Step 5: The solution set is $(1, 2)$. To check, substitute the solution into both of the original equations.

Equation 1: $\qquad 3x - 2y = -1 \qquad\qquad$ Equation 2: $\qquad -4y = -x - 7$
$$3(1) - 2(2) = -1 \qquad\qquad\qquad\qquad -4(2) = -1 - 7$$
$$3 - 4 = -1 \qquad\qquad\qquad\qquad\qquad -8 = -8$$
$$-1 = -1$$

The point $(1, 2)$ is the point of intersection.

For each of the following pairs of equations, find the point of intersection by adding the two equations together.

1. $x + 2y = 8$
   $-x - 3y = 2$

2. $x - y = 5$
   $2x + y = 1$

3. $x - y = -1$
   $x + y = 9$

4. $3x - y = -1$
   $x + y = 13$

5. $-x + 4y = 2$
   $x + y = 8$

6. $x + 4y = 10$
   $x + 7y = 16$

7. $2x - y = 2$
   $4x - 9y = -3$

8. $x + 3y = 13$
   $5x - y = 1$

9. $-x = y - 1$
   $x = y - 1$

10. $x - y = 2$
    $2y + x = 5$

11. $5x + 2y = 1$
    $4x + 8y = 20$

12. $3x - 2y = 14$
    $x - y = 6$

13. $2x + 3y = 3$
    $3x + 5y = 5$

14. $x - 4y = 6$
    $-x - y = -1$

15. $x = 2y + 3$
    $y = 3 - x$

Copyright © American Book Company

181

Chapter 15   Systems of Equations

15.7   Solving Word Problems with Systems of Equations

Certain word problems can be solved using systems of equations.

Example 8:   In a game show, Andre earns 6 points for every right answer and loses 12 points for every wrong answer. He has answered correctly 12 times as many as he has missed. His final score was 120. How many times did he answer correctly?

Step 1:   Let r = number of right answers. Let w = number of wrong answers.
We know 2 sets of information that can be made into equations with 2 variables.
He earns $+6$ points for right answers and loses 12 points for wrong answers.

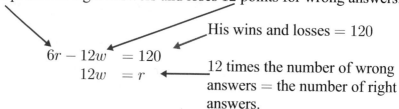

$$6r - 12w = 120$$
$$12w = r$$

His wins and losses = 120

12 times the number of wrong answers = the number of right answers.

Step 2:   Substitute the value for r ($12w$) in the first equation.
$$6(12w) - 12w = 120$$
$$w = 2$$

Step 3:   Substitute the value for w back in the equation.
$$6r - 12(2) = 120$$
$$r = 24$$

Example 9:   Ms. Sudberry bought pencils and stickers for her first grade class on two different days. The pencils and stickers cost the same each time she went to the store. How much did she pay for each pencil?

	Pencils	Stickers	Total Cost
Tuesday	30	40	$47.50
Saturday	60	5	$20.00

Step 1:   Set up your two equations. Let the price of pencils equal x, and the price of stickers equal y.
The amount of the pencils times the price of pencils (x) plus the amount of the stickers times the price of stickers (y) equals the total amount paid for both pencils and stickers.
Equation 1: $30x + 40y = \$47.50$
Equation 2: $60x + 5y = \$20.00$

Step 2:   Solve the equations by using one of the methods taught in this chapter. We will use the adding and subtracting method. First, multiply equation 1 by -2, so x will have the same coefficient in each equation but with opposite signs.
$$-2(30x + 40y = \$47.50) = -60x - 80y = -\$95.00$$

182

15.7   Solving Word Problems with Systems of Equations

Step 3:   Add the new equation 1 to equation 2.

$$
\begin{array}{rrrrl}
-60x & - & 80y & = & -\$95.00 \\
+\quad 60x & + & 5y & = & \$20.00 \\
\hline
0x & - & 75y & = & -\$75.00
\end{array}
$$

The new equation is $-75y = -\$75.00$.

Step 4:   Solve for y.
$-75y = -\$75.00$
$y = \$1.00$
Now, we know the price of stickers, but the question asked for the price of each pencil.

Step 5:   Substitute the value of y into either equation and solve for x to find the price of each pencil.
$30x + 40y = \$47.50$
$30x + 40\,(\$1.00) = \$47.50$
$30x + \$40.00 = \$47.50$
$30x = \$7.50$
$x = \$0.25$
The cost of each pencil is $\$0.25$.

Use systems of equations to solve the following word problems.

1. The sum of two numbers is 150 and their difference is 30. What are the two numbers?

2. The sum of two numbers is 121 and their difference is 37. What are the two numbers?

3. Kayla gets paid $10.00 for raking leaves and $15.00 for mowing the lawn of each of the neighbors in her subdivision. This year she mowed the lawns 10 times more than she raked leaves. In total, she made $800.00 for doing both. How many times did she rake the leaves?

4. Prices for the movie are $6.50 for children and $9.00 for adults. The total amount of ticket sales is $1,215.50. There are 152 tickets sold. How many adults and children buy tickets?

5. A farmer sells a dozen eggs at the market for $1.50 and one of his bags of grain for $6.00. He has sold 3 times as many bags of grain as he has dozens of eggs. By the end of the day, he has made $351.00 worth of sales. How many bags of grain did he sell?

6. Every time Lauren does one of her chores, she gets 20 minutes to talk on the phone. When she does not perform one of her chores, she gets 30 minutes of phone time taken away. This week she has done her chores 3 times more than she has not performed her chores. In total, she has accumulated 120 minutes. How many times has Lauren not performed her chores?

Copyright © American Book Company

Chapter 15   Systems of Equations

7. The choir sold boxes of candy and teddy bears near Valentine's Day to raise money. They sold four times as many boxes of candy as they did teddy bears. Bears sold for $11.00 each and candy sold for $5.00. They collected $558. How much of each item did they sell?

8. Mr. Marlow keeps ten and twenty dollar bills in his dresser drawer. He has 4 less than twice as many tens as twenties. He has $600 altogether. How many ten dollar bills does he have?

9. Kosta is a contestant on a math quiz show. For every correct answer, Kosta receives $22.00. For every incorrect answer, Kosta loses $30.00. Kosta answers the questions correctly twice as often as he answers the questions incorrectly. In total, Kosta wins $140.00. How many questions does Kosta answer incorrectly?

10. John Vasilovik works in landscaping. He gets paid $100 for each house he pressure-washes and $25 for each lawn he mows. He gets 6 times more jobs for mowing lawns than for pressure-washing houses. During a given month, John earns $1,000. How many houses does John pressure wash?

11. Every time Stephen walks the dog, he gets 45 minutes to play video or computer games. When he does not take out the dog on time, he gets a mess to clean up and loses 1.5 hours of video or computer game time. This week he has walked the dog on time 3 times more than he did not walk the dog on time. In total, he has accumulated 135 minutes of video or computer time. How many times has Stephen not walked the dog on time?

12. On Friday, Rosa bought party hats and kazoos for her friend's birthday party. On Saturday, she decided to purchase more when she found out more people were coming. How much did she pay for each party hat?

	Hats	Kazoos	Total Cost
Friday	12	16	$17.00
Saturday	8	4	$8.00

13. Timothy and Jesse went to purchase sports clothing they needed to play soccer. The table below shows what they bought and the amount they paid. What is the price of 1 soccer jersey?

	Soccer Jerseys	Tube Socks	Total Cost
Timothy	4	6	$91.00
Jesse	3	3	$61.50

184   Copyright © American Book Company

Chapter 15 Review

Chapter 15 Review

1. What is the equation of a line that is perpendicular to the line $3x + 2y = 6$ and passes through the point $(12, -15)$?

2. What is the equation of a line that is parallel to the line $-5x + y = -4$ and passes through the point $(-1, 7)$?

For each pair of equations below, tell whether the graphs of the lines are identical, parallel, or intersecting and determine the number of solutions the systems have.

3. $y = \frac{1}{2}x + 4$
   $y = \frac{1}{2}x - 3$

5. $\frac{1}{2} = 2x - \frac{1}{2}y$
   $4x = y + 1$

7. $y - \frac{1}{8}x = 2$
   $8y = 16 + x$

4. $3 - x = y$
   $2x + y = 1$

6. $y = 5x + 5$
   $y = 1 + 5x$

8. $3x - y = 2$
   $x + y = 2$

Find the common solution for each of the following pairs of equations using substitution.

9. $2x - y = 3$
   $3x + 4y = -1$

11. $-y = -2x + 7$
    $-x = -2y - 2$

13. $2x = y - 10$
    $-\frac{1}{2}x = y$

10. $x + 2y = 1$
    $x + 3y = 1$

12. $3x + 6y = 18$
    $4y - 2x = 12$

14. $-2x + y = -6$
    $2x - 2y = 16$

Find the point of intersection for each pair of equations by adding and/or subtracting the two equations.

15. $2x + y = 4$
    $3x - y = 6$

17. $x + y = 1$
    $y = x + 7$

19. $2x - 2y = 7$
    $3x - 5y = \frac{5}{2}$

16. $x + 2y = 3$
    $x + 5y = 0$

18. $2x + 4y = 5$
    $3x + 8y = 9$

20. $x - 3y = -2$
    $y = -\frac{1}{3}x + 4$

Use systems of equations to solve the following word problems.

21. Hargrove High School sold 381 tickets for their last basketball game. Adult tickets sold for $7 and student tickets were $3. How many adult tickets were sold if the ticket sales totaled $1,527$?

22. Zack is an ostrich and llama breeder. He sells full-grown ostriches for $925 and full-grown llamas for $890 each. Zack sold 1 less than twice as many llamas as ostriches this year. His total sales for the year were $23,455.00. How many llamas did Zack sell during this year?

23. Sarah and Abdul played Geography Quiz Bowl during summer school. For every time Abdul got an answer right, Sarah got 3 answers right. If Sarah and Abdul correctly answered 92 questions, how many times did Abdul answer correctly?

Copyright © American Book Company

185

Chapter 15   Systems of Equations

Chapter 15 Test

1. Leon graphs the function $y = 5x - 2$. Which equation would have a graph perpendicular to Leon's graph?

   A. $y = \frac{1}{5}x - 9$

   B. $y = -5x + 2$

   C. $y = 5x + 1$

   D. $y = -\frac{1}{5}x + 4$

   E. $y = 5x + 4$

2. The graph of which pair of equations below will be parallel?

   F. $x + 4y = 3$
   $3x + 4y = 3$

   G. $x - 4y = 3$
   $4y - 2x = -3$

   H. $2x - 8 = 2y$
   $2x + 8 = 2y$

   J. $6x + 6 = 6y$
   $11x - 12 = 7y$

   K. $x + 6 = 6y$
   $x - 7 = 7y$

3. Consider the following equations:

   $f(x) = 6x + 2$ and $f(x) = 3x + 2$

   Which of the following statements is true concerning the graphs of these equations?

   A. The lines are identical.

   B. The lines intersect at exactly one point.

   C. The lines are parallel to each other.

   D. The graphs of the lines intersect each other at the point $(2, 2)$.

   E. None of the above.

4. The graphs of the equations $2x + 2y = 4$ and $5x + 5y = 10$ are

   F. identical.

   G. parallel.

   H. intersecting.

   J. perpendicular.

   K. none of the above.

5. What is the solution to the equations $y = 4x - 8$ and $y = 2x$?

   A. $(-4, -8)$

   B. $(4, 2)$

   C. $(-1, -2)$

   D. $(1, 2)$

   E. $(4, 8)$

6. Two numbers have a sum of 166 and a difference of 58. What are the two numbers?

   F. 89 and 77

   G. 49 and 117

   H. 65 and 101

   J. 53 and 115

   K. 54 and 112

186   Copyright © American Book Company

Chapter 15 Test

7. Two lines are shown on the grid. One line passes through the origin and the other passes through $(-1, -1)$ with a y-intercept of 2. Which pair of equations below the grid identifies these lines?

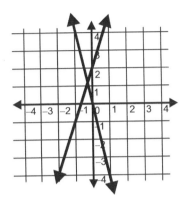

A. $y = \frac{1}{4}x$ and $y = \frac{1}{3}x + 2$

B. $x - 2y = 6$ and $4x + y = 4$

C. $y = 4x$ and $y = \frac{1}{3}x$

D. $y = 3x + 2$ and $y = -4x$

E. $y = 4x$ and $y = \frac{1}{3}x + 2$

8. Which ordered pair is a solution for the following system of equations?

$$-3x + 7y = 25$$
$$3x + 3y = -15$$

F. $(-13, -2)$

G. $(-6, 1)$

H. $(-3, -2)$

J. $(-20, -5)$

K. $(1, 3)$

9. For the following pair of equations, find the point of intersection (common solution) using the substitution method.

$$-3x - y = -2$$
$$5x + 2y = 20$$

A. $(2, -4)$

B. $(2, 5)$

C. $(-16, 50)$

D. $\left(\frac{1}{5}, \frac{1}{2}\right)$

E. $(1, 3)$

Chapter 16
Angles

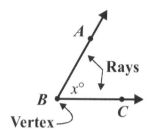

Angles are made up of two rays with a common endpoint. Rays are named by the endpoint B and another point on the ray. Ray $\overrightarrow{BA}$ and ray $\overrightarrow{BC}$ share a common endpoint.

Angles are usually named by three capital letters. The middle letter names the vertex. The angle to the left can be named $\angle ABC$ or $\angle CBA$. An angle can also be named by a lower case letter between the sides, $\angle x$, or by the vertex alone, $\angle B$.

A protractor, ⌒, is used to measure angles. The protractor is divided evenly into a half circle of 180 degrees (180°). When the middle of the bottom of the protractor is placed on the vertex, and one of the rays of the angle is lined up with 0°, the other ray of the angle crosses the protractor at the measure of the angle. The angle below has the ray pointing left lined up with 0° (the outside numbers), and the other ray of the angle crosses the protractor at 55°. The angle measures 55°.

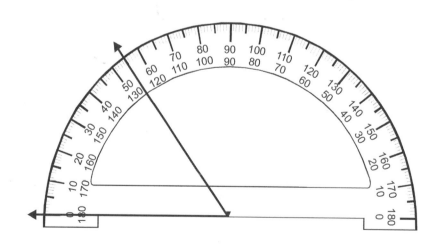

16.1 Types of Angles

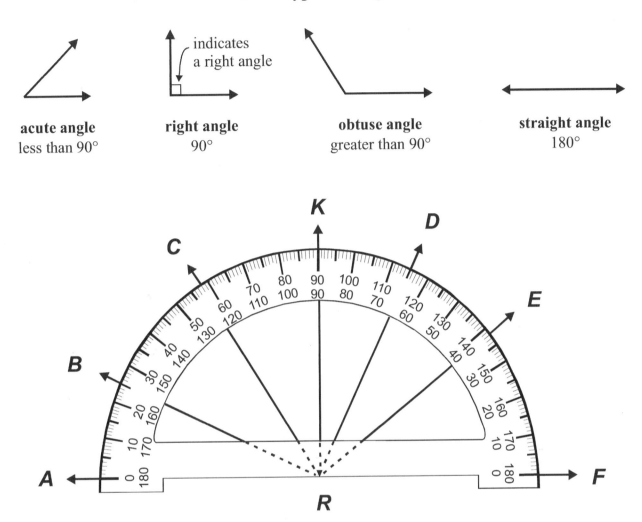

Using the protractor above, find the measure of the following angles. Then, tell what type of angle it is: acute, right, obtuse, or straight.

		Measure	Type of Angle
1.	What is the measure of angle ARF?	_____	_____
2.	What is the measure of angle CRF?	_____	_____
3.	What is the measure of angle BRF?	_____	_____
4.	What is the measure of angle ERF?	_____	_____
5.	What is the measure of angle ARB?	_____	_____
6.	What is the measure of angle KRA?	_____	_____
7.	What is the measure of angle CRA?	_____	_____
8.	What is the measure of angle DRF?	_____	_____
9.	What is the measure of angle ARD?	_____	_____
10.	What is the measure of angle FRK?	_____	_____

Chapter 16 Angles

16.2 Adjacent Angles

Adjacent angles are two angles that have the same vertex and share one ray. They do not share space inside the angles.

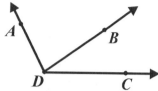

∠ADB is **adjacent** to ∠BDC. However, ∠ADB is **not adjacent** to ∠ADC because adjacent angles do not share any space inside the angle

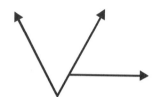

These two angles are **not adjacent**. They share a common ray but do not share the same vertex.

For each diagram below, name the angle that is adjacent to it.

1.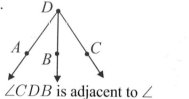
∠CDB is adjacent to ∠_____

5.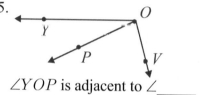
∠YOP is adjacent to ∠_____

2.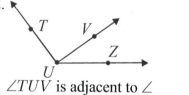
∠TUV is adjacent to ∠_____

6.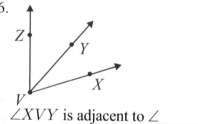
∠XVY is adjacent to ∠_____

3.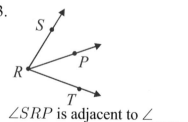
∠SRP is adjacent to ∠_____

7.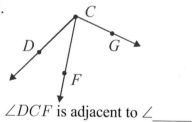
∠DCF is adjacent to ∠_____

4.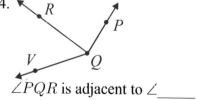
∠PQR is adjacent to ∠_____

8.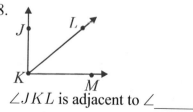
∠JKL is adjacent to ∠_____

16.3 Vertical Angles

When two lines intersect, two pairs of vertical angles are formed. Vertical angles are not adjacent. Vertical angles have the same measure.

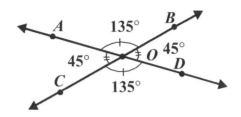

$\angle AOB$ and $\angle COD$ are vertical angles. $\angle AOC$ and $\angle BOD$ are vertical angles. **Vertical angles are congruent.** Congruent means they have the same measure.

In the diagram below, name the second angle in each pair of vertical angles.

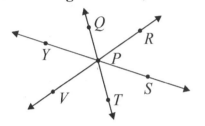

1. $\angle YPV$ _____
2. $\angle QPR$ _____
3. $\angle SPT$ _____
4. $\angle VPT$ _____
5. $\angle RPT$ _____
6. $\angle VPS$ _____
7. $\angle MLN$ _____
8. $\angle KLH$ _____
9. $\angle GLN$ _____
10. $\angle GLM$ _____
11. $\angle KLM$ _____
12. $\angle HLG$ _____

Use the information given to find the measure of each unknown vertical angle.

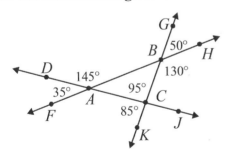
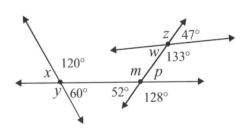

13. $\angle CAF =$ _____
14. $\angle ABC =$ _____
15. $\angle KCJ =$ _____
16. $\angle ABG =$ _____
17. $\angle BCJ =$ _____
18. $\angle CAB =$ _____

19. $\angle x =$ _____
20. $\angle y =$ _____
21. $\angle z =$ _____
22. $\angle w =$ _____
23. $\angle m =$ _____
24. $\angle p =$ _____

Chapter 16 Angles

16.4 Complementary and Supplementary Angles

Two angles are **complementary** if the sum of the measures of the angles is 90°.

Two angles are **supplementary** if the sum of the measures of the angles is 180°. A **linear pair** is a pair of adjacent angles that are supplementary. Below the angles 32° and 148° are a linear pair. The angles may be adjacent but do not need to be.

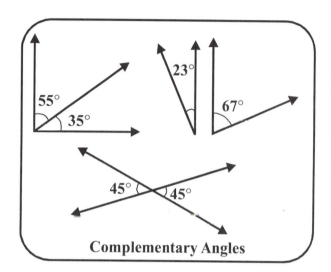
Complementary Angles

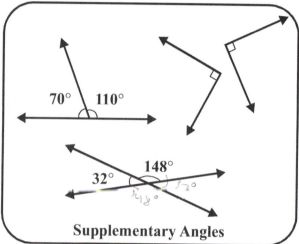
Supplementary Angles

Calculate the measure of each unknown angle.

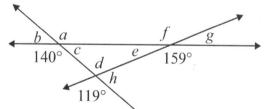

1. ∠a = _____
2. ∠b = _____
3. ∠c = _____
4. ∠d = _____
5. ∠e = _____
6. ∠f = _____
7. ∠g = _____
8. ∠h = _____

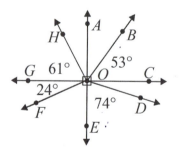
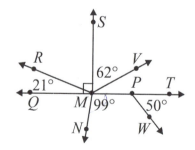
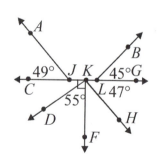

9. ∠AOB = _____
10. ∠COD = _____
11. ∠EOF = _____
12. ∠AOH = _____

13. ∠RMS = _____
14. ∠VMT = _____
15. ∠QMN = _____
16. ∠WPQ = _____

17. ∠AJK = _____
18. ∠CKD = _____
19. ∠FKH = _____
20. ∠BLC = _____

16.5 Corresponding, Alternate Interior, and Alternate Exterior Angles

If two parallel lines are intersected by a **transversal**, a line passing through both parallel lines, the **corresponding angles** are congruent.

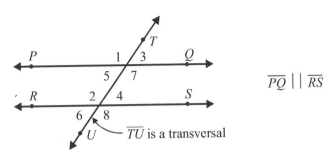

$\overline{PQ} \parallel \overline{RS}$

$\overline{TU}$ is a transversal

∠1 and ∠2 are congruent.   They are corresponding angles.
∠3 and ∠4 are congruent.   They are corresponding angles.
∠5 and ∠6 are congruent.   They are corresponding angles.
∠7 and ∠8 are congruent.   They are corresponding angles.

Alternate interior angles are also congruent. They are on the opposite sides of the transversal and inside the parallel lines.

∠5 and ∠4 are congruent.   They are alternate interior angles.
∠7 and ∠2 are congruent.   They are alternate interior angles.

Alternate exterior angles are also congruent. They are on the opposite sides of the transversal and above and below the parallel lines.

∠1 and ∠8 are congruent.   They are alternate exterior angles.
∠3 and ∠6 are congruent.   They are alternate exterior angles.

Look at the diagram below. For each pair of angles, state whether they are corresponding (C), alternate interior (I), alternate exterior (E), vertical (V), or supplementary angles (S).

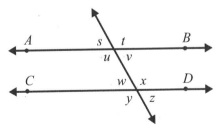

$\overline{AB}$ and $\overline{CD}$ are parallel.

1. ∠u, ∠x
2. ∠w, ∠s
3. ∠t, ∠y
4. ∠s, ∠t
5. ∠w, ∠y

6. ∠t, ∠x
7. ∠w, ∠z
8. ∠v, ∠w
9. ∠v, ∠z
10. ∠s, ∠z

11. ∠t, ∠u
12. ∠w, ∠x
13. ∠w, ∠s
14. ∠s, ∠v
15. ∠x, ∠z

Chapter 16 Angles

16.6 Sum of Interior Angles of a Polygon

Given a convex polygon, you can find the sum of the measures of the interior angles using the following formula: Sum of the measures of the interior angles $= 180°(n-2)$, where n is the number of sides of the polygon.

Example 1: Find the sum of the measures of the interior angles of the following polygon:

Solution: The figure has 8 sides. Using the formula we have $180°(8-2) = 180°(6) = 1080°$

Using the formula, $180°(n-2)$, find the sum of the interior angles of the following figures.

1.
4.
7.
10.

2.
5.
8.
11.

3.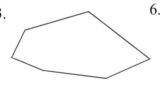
6.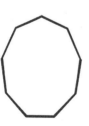
9.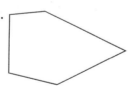
12.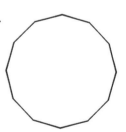

Find the measure of $\angle G$ in the regular polygons shown below. Remember that the sides of a regular polygon are equal.

13.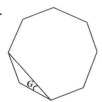
14.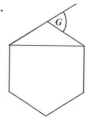
15.

Chapter 16 Review

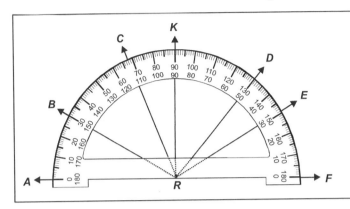

1. What is the measure of ∠DRA?

2. What is the measure of ∠CRF?

3. What is the measure of ∠ARB?

Use the following diagram for questions 4–14.

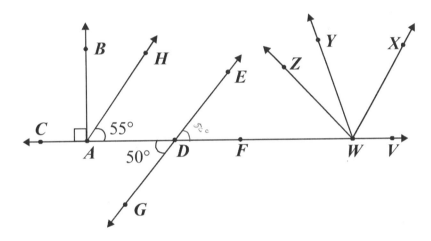

4. Which angle is a supplementary angle to ∠EDF?

5. What is the measure of ∠GDF?

6. Which two angles are right angles?

7. What is the measure of ∠EDF?

8. Which angle is adjacent to ∠BAD?

9. Which angle is a complementary angle to ∠HAD?

10. What is the measure of ∠HAB?

11. What is the measure of ∠CAD?

12. What kind of angle is ∠FDA?

13. What kind of angle is ∠GDA?

14. Which angles are adjacent to ∠EDA?

Chapter 16 Angles

Look at the diagram below. For each pair of angles, state whether they are corresponding (C), alternate interior (I), alternate exterior (E), vertical (V), or supplementary (S) angles.

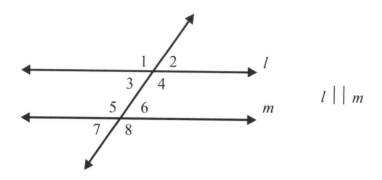

15. ∠1 and ∠4

16. ∠2 and ∠6

17. ∠1 and ∠3

18. ∠5 and ∠8

19. ∠5 and ∠7

20. ∠6 and ∠5

21. ∠2 and ∠7

22. ∠1 and ∠2

23. ∠4 and ∠5

24. ∠6 and ∠8

25. ∠3 and ∠6

26. ∠4 and ∠8

27. ∠1 and ∠5

28. ∠2 and ∠3

29. What is the sum of the measures of the interior angles in the figure below?

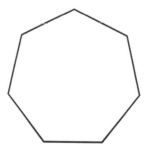

Chapter 16 Test

1. What type of angle is shown below?

   A. right
   B. acute
   C. obtuse
   D. straight
   E. perpendicular

2. What is the sum of two complementary angles?

   F. 180°
   G. 45°
   H. 270°
   J. 360°
   K. 90°

3. What is the measure of an angle that is supplementary to 87°?

   A. −42°
   B. 3°
   C. 273°
   D. 93°
   E. 13°

4. In the diagram below, which two angles form a linear pair?

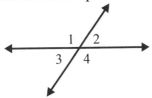

   F. ∠1 and ∠2
   G. ∠1 and ∠3
   H. ∠1 and ∠4
   J. Both F and G are correct.
   K. ∠2 and ∠3

Use the following diagram to answer questions 5–8.

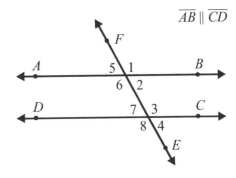

5. Which angles are vertical angles?

   A. ∠1 and ∠2
   B. ∠1 and ∠3
   C. ∠1 and ∠4
   D. ∠1 and ∠5
   E. ∠1 and ∠6

6. Which angles are alternate exterior angles?

   F. ∠2 and ∠7
   G. ∠3 and ∠8
   H. ∠1 and ∠8
   J. ∠5 and ∠3
   K. ∠6 and ∠7

7. Which angles are alternate interior angles?

   A. ∠7 and ∠8
   B. ∠6 and ∠3
   C. ∠2 and ∠3
   D. ∠1 and ∠4
   E. ∠5 and ∠8

8. Which angles are corresponding angles?

   F. ∠1 and ∠3
   G. ∠1 and ∠4
   H. ∠7 and ∠6
   J. ∠1 and ∠8
   K. ∠8 and ∠4

Chapter 17
Triangles

17.1 Types of Triangles

right triangle
contains 1 right ∠

acute triangle
all angles are acute
(less than 90°)

obtuse triangle
one angle is obtuse
(greater than 90°)

isosceles triangle
two sides equal
two angles equal

equilateral triangle
all three sides equal
all angles are 60°

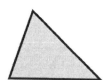

scalene triangle
no sides equal
no angles equal

17.2 Interior Angles of a Triangle

The three interior angles of a triangle always add up to $180°$.

Example 1:

$45° + 45° + 90° = 180°$ $30° + 60° + 90° = 180°$ $60° + 60° + 60° = 180°$

Example 2:   Find the missing angle in the triangle.

Solution:
$$20° + 125° + x = 180°$$
$$-20° -125° -20° -125°$$
$$x = 180° - 20° - 125°$$
$$x = 35°$$

Subtract $20°$ and $125°$ from both sides to get x by itself.

The missing angle is $35°$.

Find the missing angle in the triangles.

1.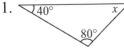

2.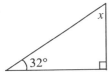

3.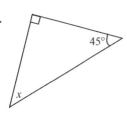

4.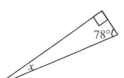

5.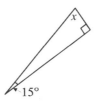

6.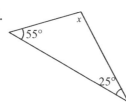

7.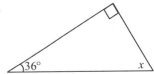

8.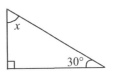

9.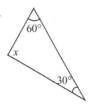

Find the missing angles in the triangles.

10.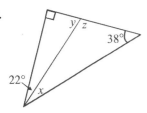

11.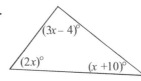

12.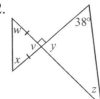

Chapter 17 Triangles

17.3 Exterior Angles

The **exterior angle** of a triangle is always equal to the sum of the opposite interior angles.

Example 3: Find the measure of $\angle x$ and $\angle y$.

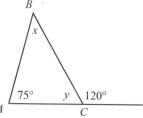

Step 1: Using the rule for exterior angles,
$120° = \angle A + \angle B$
$120° = 75° + x$
$120° - 75° = 75° - 75° + x$
$45° = x$

Step 2: The sum of the interior angles of a triangle equals $180°$, so
$180° = 75° + 45° + y$
$180° - 75° - 45° = 75° - 75° + 45° - 45° + y$
$60° = y$

Find the measures of x and y.

1.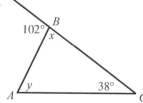

3.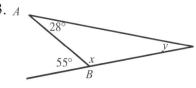

5.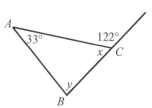

2.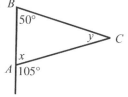

4.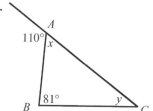

6.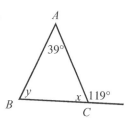

Find the measures of the angles.

7.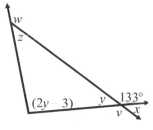

8.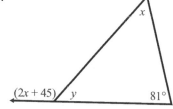

9.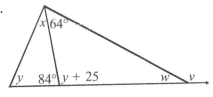

17.4 Triangle Inequality Theorem

The triangle inequality theorem states that the sum of the measure of any two sides in a triangle must be greater than the measure of the third side.

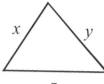

$x + y > z$
$y + z > x$
$x + z > y$

Example 4: Determine whether or not it is possible to create a triangle with sides of 1 units, 5 units, and 7 units.

Step 1: First, you must set up three inequalities. Remember the sum of any two sides of a triangle must be greater than the third side.

$1 + 5 > 7$ $1 + 7 > 5$ $5 + 7 > 1$

Step 2: Determine if the inequalities are true.

$1 + 5 > 7$ $1 + 7 > 5$ $5 + 7 > 1$
$6 > 7$ $8 > 5$ $12 > 1$
False   True   True

The number 6 is not greater than 7, so a triangle cannot be formed using the sides given.
(All three inequalities must be true in order to create a triangle.)

Determine whether or not it is possible to create a triangle given the following measures of sides. Write "yes" if it is possible to form a triangle with the given measures of sides or write "no" if it is not possible.

1. 7, 8, 13
2. 2, 5, 9
3. 10, 8, 15
4. 6, 9, 20
5. 101, 89, 150
6. 1, 2, 4
7. 7, 7, 14
8. 21, 15, 29
9. 11, 9, 17

Chapter 17   Triangles

17.5   Similar Triangles

Two triangles are similar if the measurements of the three angles in both triangles are the same. If the three angles are the same, then their corresponding sides are proportional.

Corresponding Sides - The triangles below are similar. Therefore, the two shortest sides from each triangle, c and f, are corresponding. The two longest sides from each triangle, a and d, are corresponding. The two medium length sides, b and e, are corresponding.

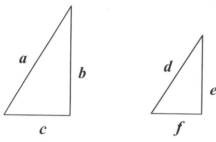

Proportional - The corresponding sides of similar triangles are proportional to each other. This means if we know all the measurements of one triangle, and we know only one measurement of the other triangle, we can figure out the measurements of the two other sides with proportion problems. The two triangles below are similar.

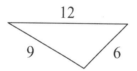   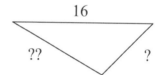

Note: To set up the proportion correctly, it is important to keep the measurements of each triangle on opposite sides of the equal sign.

To find the short side:	To find the medium length side:
Step 1: Set up the proportion	**Step 1:** Set up the proportion
$\dfrac{\text{long side}}{\text{short side}} \quad \dfrac{12}{6} = \dfrac{16}{?}$	$\dfrac{\text{long side}}{\text{medium}} \quad \dfrac{12}{9} = \dfrac{16}{??}$
Step 2: Solve the proportion. Multiply the two numbers diagonal to each other and then divide by the other number.	**Step 2:** Solve the proportion. Multiply the two numbers diagonal to each other and then divide by the other number.
$16 \times 6 = 96$	$16 \times 9 = 144$
$96 \div 12 = 8$	$144 \div 12 = 12$

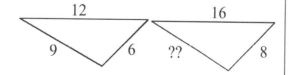   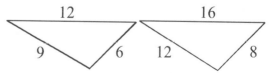

17.5 Similar Triangles

To find the scale factor in the problem on the previous page, we must divide a value from the second triangle by the corresponding value from the first triangle. The value 16 is from the second triangle, and the corresponding value from the first triangle is 12. $k = \dfrac{16}{12} = \dfrac{4}{3}$

The scale factor in this problem is $\frac{4}{3}$.

To check this answer multiply every term in the first triangle by the scale factor, and you will get every term in the second triangle.

$$12 \times \frac{4}{3} = 16 \qquad 9 \times \frac{4}{3} = 12 \qquad 6 \times \frac{4}{3} = 8$$

Find the missing side from the following similar triangles.

1.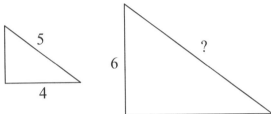

5.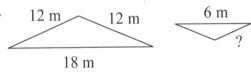

2.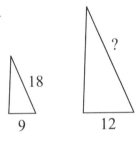

6.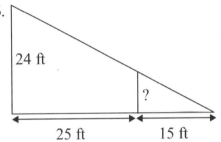

3.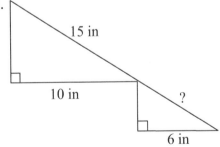

7.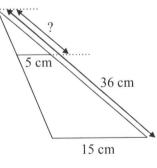

4.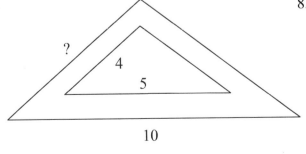

8.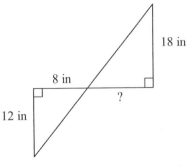

Copyright © American Book Company

Chapter 17 Triangles

17.6 Pythagorean Theorem

Pythagoras was a Greek mathematician and philosopher who lived around 600 B.C. He started a math club among Greek aristocrats called the Pythagoreans. Pythagoras formulated the **Pythagorean Theorem** which states that in a **right triangle**, the sum of the squares of the legs of the triangle are equal to the square of the hypotenuse. Most often you will see this formula written as $a^2 + b^2 = c^2$. **This relationship is only true for right triangles.** **Pythagorean Triples** are the set of three positive integers which satisfies the Pythagorean theorem.

Example 5: Find the length of side c.

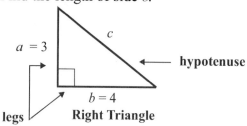

Formula: $a^2 + b^2 = c^2$
$3^2 + 4^2 = c^2$
$9 + 16 = c^2$
$25 = c^2$
$\sqrt{25} = \sqrt{c^2}$
$5 = c$

Find the hypotenuse of the following triangles. Round the answers to two decimal places.

1.

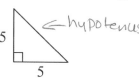
$c = $ _____

2.
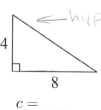
$c = $ _____

3.
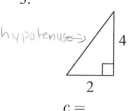
$c = $ _____

4.
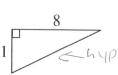
$c = $ _____

5.
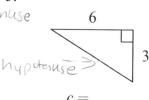
$c = $ _____

6.
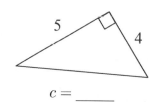
$c = $ _____

7.
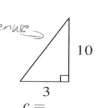
$c = $ _____

8.
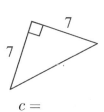
$c = $ _____

9.
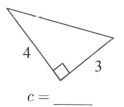
$c = $ _____

17.7 Finding the Missing Leg of a Right Triangle

In some triangles, we know the measurement of the hypotenuse as well as one of the legs. To find the measurement of the other leg, use the Pythagorean theorem by filling in the known measurements, and then solve for the unknown side.

Example 6: Find the measure of b.

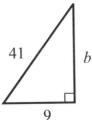

In the formula, $a^2 + b^2 = c^2$, a and b are the legs and c is always the hypotenuse.

$9^2 + b^2 = 41^2$
$81 + b^2 = 1681$
$b^2 = 1681 - 81$
$b^2 = 1600$
$\sqrt{b^2} = \sqrt{1600}$
$b = 40$

Practice finding the measure of the missing leg in each right triangle below. Simplify square roots.

1.
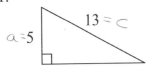

2.
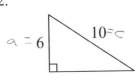

3.
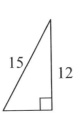

4.
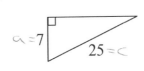

5.
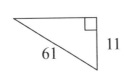

6.
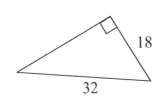

7.

8.
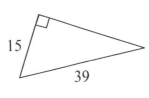

9.
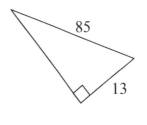

Copyright © American Book Company

Chapter 17 Triangles

17.8 Applications of the Pythagorean Theorem

The Pythagorean Theorem can be used to determine the distance between two points in some situations. Recall that the formula is written $a^2 + b^2 = c^2$.

Example 7: Find the distance between point B and point A given that the length of each square is 1 inch long and 1 inch wide.

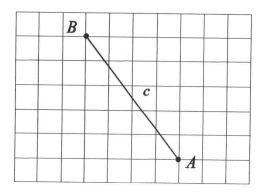

Step 1: Draw a straight line between the two points. We will call this side c.

Step 2: Draw two more lines, one from point B and one from point A. These lines should make a $90°$ angle. The two new lines will be labeled a and b. Now we can use the Pythagorean Theorem to find the distance from Point B to Point A.

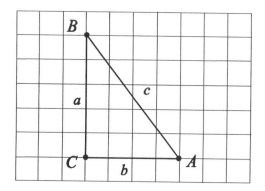

Step 3: Find the length of a and c by counting the number of squares each line has. We find that $a = 5$ inches and $b = 4$ inches. Now, substitute the values found into the Pythagorean Theorem.

$$\begin{aligned} a^2 + b^2 &= c^2 \\ 5^2 + 4^2 &= c^2 \\ 25 + 16 &= c^2 \\ 41 &= c^2 \\ \sqrt{41} &= \sqrt{c^2} \\ \sqrt{41} &= c \end{aligned}$$

17.8 Applications of the Pythagorean Theorem

Use the Pythagorean Theorem to find the distances asked. Round your answers to two decimal points.

Below is a diagram of the mall. Use the grid to help answer questions 1 and 2. Each square is 25 feet × 25 feet.

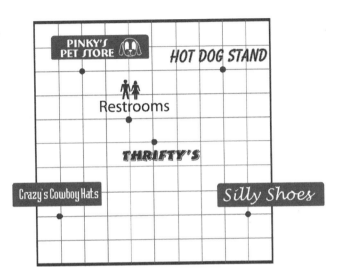

1. Marty walks from Pinky's Pet Store to the restroom to wash his hands. How far did he walk?

2. Betty needs to meet her friend at Silly Shoes, but she wants to get a hot dog first. If Betty is at Thrifty's, how far will she walk to meet her friend?

Below is a diagram of a football field. Use the grid on the football field to help find the answers to questions 3 and 4. Each square is 10 yards × 10 yards.

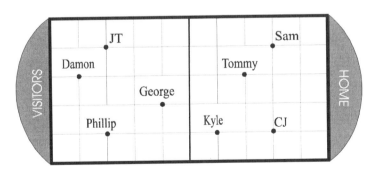

3. George must throw the football to a teammate before he is tackled. If CJ is the only person open, how far must George be able to throw the ball?

4. Damon has the football and is close to scoring a touchdown. If Phillip tries to stop him, how far must he run to reach Damon?

Chapter 17 Review

1. Find the missing angle.

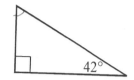

2. What is the length of line segment $\overline{WY}$?

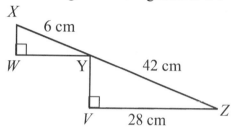

3. Find the missing side.

4. Find the measure of the missing leg of the right triangle below.

5. The following two triangles are similar. Find the length of the missing side.

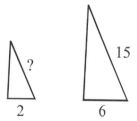

For questions 6–7, determine if the measures of sides given can form a triangle.

6. 1, 5, 3   x+y>z
             y+z>x
7. 16, 22, 31   x+z>y

8.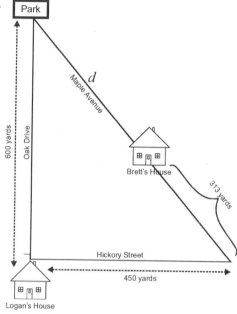

Logan enjoys taking his dog to the park. Some days he leaves his house, located on the corner of Hickory St. and Oak Dr., and walks directly to the park. Sometimes, though, he walks down Hickory St., turns onto Maple Ave. to meet his friend, Brett, and then continues on Maple Ave. to the park. What is the approximate distance (d) from Brett's house to the park?

For questions 9–10, find the missing angles.

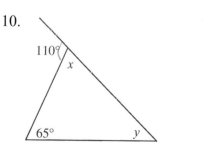

Chapter 17 Test

1. What is the measure of missing angle?

   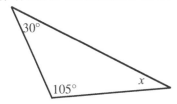

   A. 225°
   B. 40°
   C. 75°
   D. 30°
   E. 45°

2. What is the measure of y?

   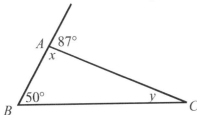

   F. 93°
   G. 87°
   H. 37°
   J. 47°
   K. Cannot be determined

3. What is the measure of the missing side in the triangle?

   A. 5 cm
   B. 29 cm
   C. 10 cm
   D. 20 cm
   E. $\sqrt{15}$ cm

4. Given the measures of sides below, which cannot form a triangle?

   F. 5, 7, 10
   G. 2, 3, 4
   H. 15, 6, 9
   J. 19, 20, 36
   K. 12, 13, 12

5. What type of triangle is illustrated?

   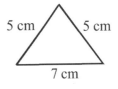

   A. right
   B. isosceles
   C. equilateral
   D. obtuse
   E. scalene

6. Approximately what is the measure of the hypotenuse of the triangle?

   F. 14 in
   G. 157 in
   H. 313 in
   J. 25 in
   K. 18 in

7. What is the measure of the two missing angles?

   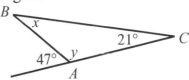

   A. $x = 26°, y = 133°$
   B. $x = 47°, y = 112°$
   C. $x = 112°, y = 47°$
   D. $x = 133°, y = 26°$
   E. $x = 133°, y = 47°$

8. What is the length of the line segment $\overline{WY}$?

   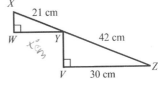

   F. 15 cm
   G. 16 cm
   H. 18 cm
   J. 30 cm
   K. 23 cm

Chapter 18
Plane Geometry

18.1 Points

A **point** is defined as a location in space that has neither length nor width. Although it is an undefined term in geometry, it usually represents something such as a star in a constellation. Points are normally represented as dots on a line or a graph, and are named with capital letters.

18.2 Lines, Segments, and Rays

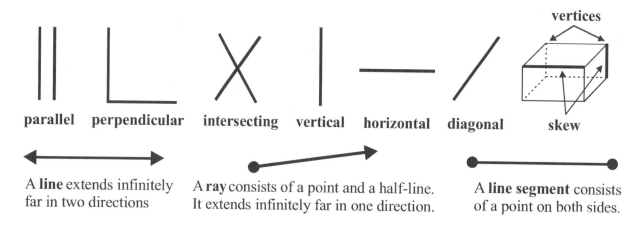

Points are said to be **collinear** if they lie on the same line. Another undefined concept in geometry is **betweenness**. If three points are collinear, then one of them can be said to be between the other two. The formula for the midpoint of a line segment can be found in the next section. Line segments are **congruent** if they are exactly the same size. Points and lines in the same plane are **coplanar**.

18.3 Types of Polygons

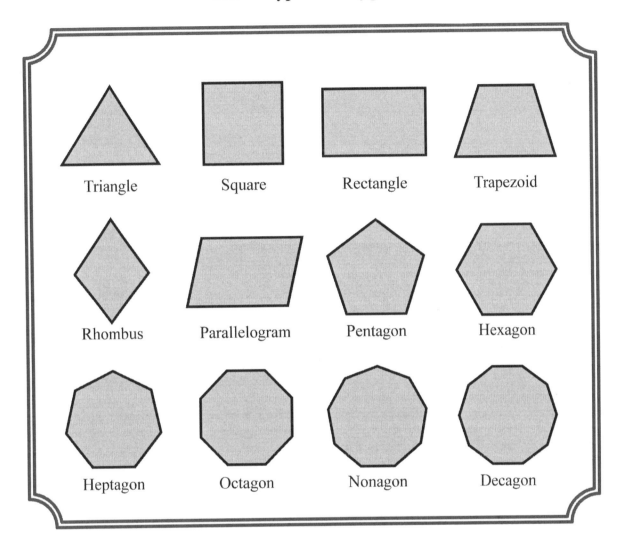

18.4 Quadrilaterals and Their Properties

A **quadrilateral** is a polygon with four sides. A **parallelogram** is a quadrilateral in which both pairs of opposite sides are parallel. The following properties of parallelograms are given without proof:

1. Both pairs of opposite sides are parallel.
2. The opposite sides are congruent.
3. The opposite angles are congruent
4. Consecutive angles are supplementary.
5. The diagonals bisect each other

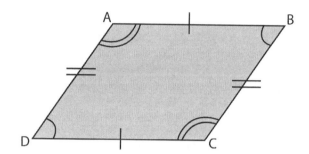

A **rectangle** is a parallelogram with four right angles. It follows that a rectangle has all of the properties listed above, plus all four angles are 90°. In addition, the diagonals of a rectangle are congruent. A **rhombus** is a parallelogram with four congruent sides. A rhombus has all the properties of a parallelogram, but both pairs of opposite sides are congruent as well as parallel. Due to this fact, the diagonals of a rhombus each bisect a pair of opposite angles and the diagonals are perpendicular to each other. A **square** is a rhombus with four right angles. Therefore, a square has four congruent sides and four congruent angles (each 90°). As you can see, a square is also a quadrilateral, a parallelogram, a rectangle, and a rhombus. These properties plus the four right angles make it a square.

A **trapezoid** is a quadrilateral with only one pair of parallel sides. The parallel sides are called bases, and the other two sides are called legs.

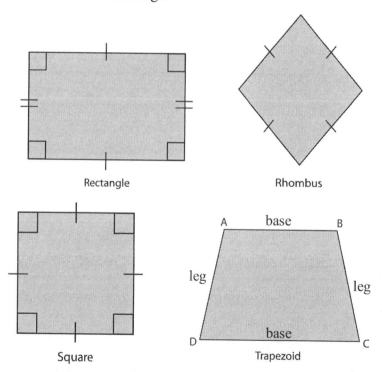

18.5 Perimeter

The **perimeter** is the distance around a polygon. To find the perimeter, add the lengths of the sides.
Examples:

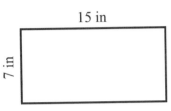

$P = 7 + 15 + 7 + 15$
$P = 44$ in

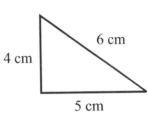

$P = 4 + 6 + 5$
$P = 15$ cm

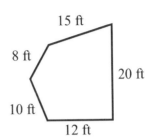

$P = 8 + 15 + 20 + 12 + 10$
$P = 65$ ft

Find the perimeter of the following polygons.

1.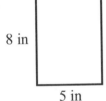

2.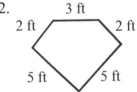

3.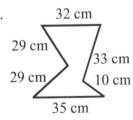

4.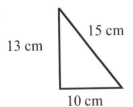

5.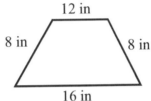

6.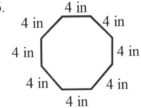

7.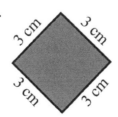

8.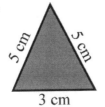

9.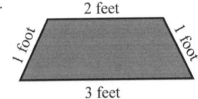

10. a regular pentagon with sides of 12 centimeters

11. a regular square with sides of 7 inches

12. a regular decagon with sides of 5 centimeters

13. a pentagon with sides of 2 feet

14. a triangle with 3 equal sides, 4 inches long

15. a regular hexagon with sides of 5 centimeters

Chapter 18 Plane Geometry

18.6 Area of Squares and Rectangles

Area - area is always expressed in square units, such as in², m², and ft².

The area, (A), of squares and rectangles equals length (l) times width (w). $A = l \times w$.

Example 1:

4 cm

4 cm

$A = lw$
$A = 4 \times 4$
$A = 16$ cm²

If a square has an area of 16 cm², it means that it will take 16 squares that are 1 cm on each side to cover the area that is 4 cm on each side.

Find the area of the following squares and rectangles using the formula $A = lw$.

1. 10 ft × 10 ft

2. 5 cm × 2 cm

3. 4 in × 9 in

4. 9 in × 20 in

5. 6 ft × 6 ft

6. 10 cm × 5 cm

7. 4 ft × 2 ft

8. 5 in × 8 in

9. 12 ft × 12 ft

10. 7 cm × 12 cm

11. 1 ft × 8 ft

12. 6 cm × 7 cm

18.7 Area of Triangles

Example 2: Find the area of the following triangle.
The formula for the area of a triangle is as follows:

$$A = \frac{1}{2} \times b \times h$$

$A =$ area
$b =$ base
$h =$ height or altitude

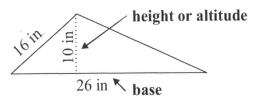

Step 1: Insert the measurements from the triangle into the formula: $A = \frac{1}{2} \times 26 \times 10$

Step 2: Cancel and multiply. $A = \frac{1}{26} \times \frac{\cancel{26}^{13}}{1} \times \frac{10}{1} = 130 \text{ in}^2$

Note: **Area is always expressed in square units such as** in^2, ft^2, **or** m^2.

Find the area of the following triangles. Remember to include units.

1.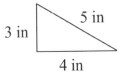

2.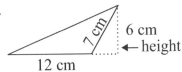

3.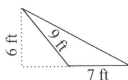

4.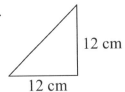

5.

6.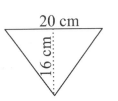

7.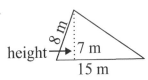

8.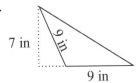

9.

10.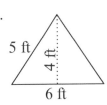

11.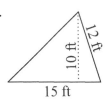

12.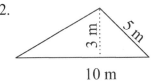

Chapter 18   Plane Geometry

18.8   Area of Trapezoids and Parallelograms

Example 3:   Find the area of the following parallelogram.

The formula for the area of a parallelogram is $A = bh$.
A = area
b = base
h = height

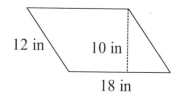

Step 1:   Insert measurements from the parallelogram into the formula: $A = 18 \times 10$.
Step 2:   Multiply. $18 \times 10 = 180$ in^2

Example 4:   Find the area of the following trapezoid.
The formula for the area of a trapezoid is $A = \frac{1}{2}h(b_1 + b_2)$. A trapezoid has two bases that are parallel to each other. When you add the length of the two bases together and then multiply by $\frac{1}{2}$, you find their average length.

A = area
b = base
h = height

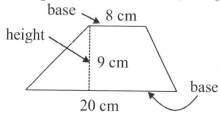

Insert the measurements from the trapezoid into the formula and solve:
$\frac{1}{2} \times 9\,(8 + 20) = 126$ cm^2

Find the area of the following parallelograms and trapezoids.

1.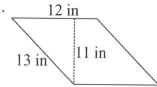

4.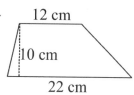

7.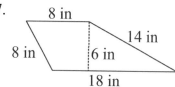

2.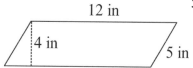

5.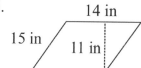

8.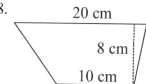

3.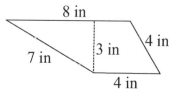

6.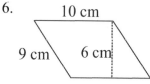

9.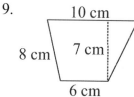

18.9 Circumference

Circumference, C - the distance around the outside of a circle
Diameter, d - a line segment passing through the center of a circle from one side to the other
Radius, r - a line segment from the center of a circle to the edge of a circle
Pi, π - the ratio of a circumference of a circle to its diameter $\pi \approx 3.14$ or $\pi \approx \frac{22}{7}$

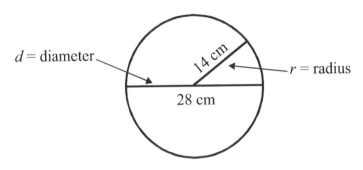

The formula for the circumference of a circle is $C = 2\pi r$ or $C = \pi d$. (The formulas are equal because the diameter is equal to twice the radius, $d = 2r$.)

Example 5: Find the approximate circumference of the circle above.

$C = \pi d$ Use $\pi \approx 3.14$ $C = 2\pi r$
$C \approx 3.14 \times 28$ $C \approx 2 \times 3.14 \times 14$
$C \approx 87.92\,\text{cm}$ $C \approx 87.92\,\text{cm}$

Use the formulas given above to find the approximate circumferences of the following circles. Use $\pi \approx 3.14$.

1. 8 in

2. 14 ft

3. 2 cm

4. 6 m

5. 8 ft

$C \approx$ _____ $C \approx$ _____ $C \approx$ _____ $C \approx$ _____ $C \approx$ _____

Use the formulas given above to find the approximate circumferences of the following circles. Use $\pi \approx \frac{22}{7}$.

6. 3 ft

7. 12 in

8. 6 m

9. 5 cm

10. 16 in

$C \approx$ _____ $C \approx$ _____ $C \approx$ _____ $C \approx$ _____ $C \approx$ _____

Chapter 18 Plane Geometry

18.10 Area of a Circle

The formula for the area of a circle is $A = \pi r^2$. The area is how many square units of measure would fit inside a circle.

Example 6: Find the approximate area of the circle, using both values for π.

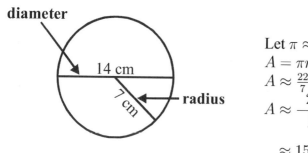

Let $\pi \approx \frac{22}{7}$
$A = \pi r^2$
$A \approx \frac{22}{7} \times 7^2$
$A \approx \frac{22}{7\,1} \times \frac{\cancel{49}\,7}{1}$
$\approx 154 \text{ cm}^2$

Let $\pi \approx 3.14$
$A = \pi r^2$
$A \approx 3.14 \times 7^2$
$A \approx 3.14 \times 49$
$\approx 153.86 \text{ cm}^2 \approx 154 \text{ cm}^2$

Find the area of the following circles. Remember to include units.

	$\pi \approx 3.14$	$\pi \approx \frac{22}{7}$
1. (5 in)	$A \approx$ ____	$A \approx$ ____
2. (16 ft)	$A \approx$ ____	$A \approx$ ____
3. (8 cm)	$A \approx$ ____	$A \approx$ ____
4. (3 m)	$A \approx$ ____	$A \approx$ ____

Fill in the chart below. Include appropriate units.

			Area	
	Radius	Diameter	$\pi \approx 3.14$	$\pi \approx \frac{22}{7}$
5.	9 ft			
6.		4 in		
7.	8 cm			
8.		20 ft		
9.	14 m			
10.		18 cm		
11.	12 ft			
12.		6 in		

218 Copyright © American Book Company

18.11 Two-Step Area Problems

Solving the problems below will require two steps. You will need to find the area of two figures, and then either add or subtract the two areas to find the answer. **Carefully read the examples.**

Example 7:
Find the area of the living room below.
Figure 1

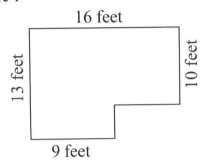

Step 1: Complete the rectangle as in Figure 2, and compute the area as if it were a complete rectangle.

Figure 2

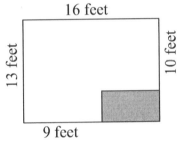

$A = \text{length} \times \text{width}$
$A = 16 \times 13$
$A = 208 \text{ ft}^2$

Step 2: Figure the area of the shaded part.

7 feet
3 feet
$7 \times 3 = 21 \text{ ft}^2$

Step 3: Subtract the area of the shaded part from the area of the complete rectangle

$208 - 21 = 187 \text{ ft}^2$

Example 8:
Find the area of the shaded sidewalk.

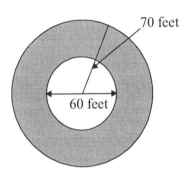

Step 1: Find the area of the outside circle.
$\pi \approx 3.14$
$A \approx 3.14 \times 70 \times 70$
$A \approx 15,386 \text{ ft}^2$

Step 2: Find the area of the inside circle.
$\pi \approx 3.14$
$A \approx 3.14 \times 30 \times 30$
$A \approx 2826 \text{ ft}^2$

Step 3: Subtract the area of the inside circle from the area of the outside circle.
$15,386 - 2826 = 12,560 \text{ ft}^2$

Chapter 18 Plane Geometry

Find the area of the following figures.

1.
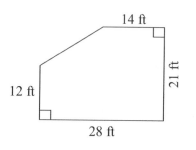

5. What is the area of the shaded part?
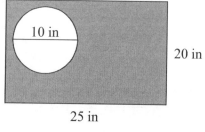

2.
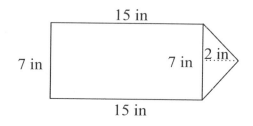

6. What is the area of the shaded part?
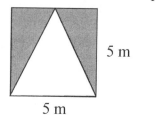

3. What is the area of the shaded circle? Use $\pi \approx 3.14$, and round the answer to the nearest whole number.
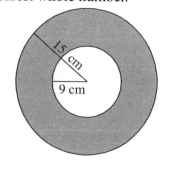

7. What is the area of the shaded part?
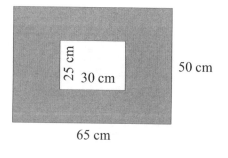

4.
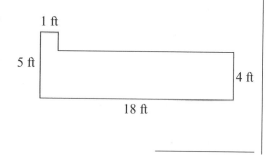

8.
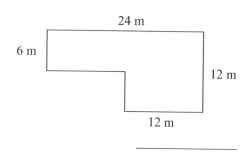

18.12 Similar Figures

The measures of corresponding sides of similar figures can also be found by setting up a proportion.

Example 9: The following rectangles are similar. Find the value of x.

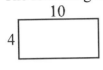
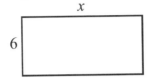

Step 1: Set up the proportion.

$$\frac{\text{Short side rectangle 1}}{\text{Long side rectangle 1}} = \frac{\text{Short side rectangle 2}}{\text{Long side rectangle 2}}$$

Step 2: $\frac{4}{10} = \frac{6}{x}$ Cross multiply.

$4x = 60$

$x = 15$

All of the pairs of figures below are similar. Find the missing side for each pair.

1.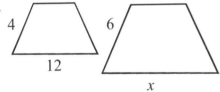

2.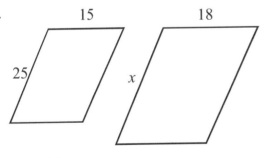

3.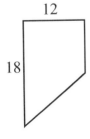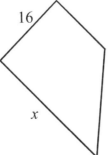

4.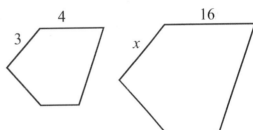

5.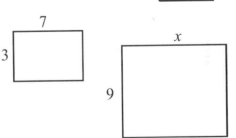

6.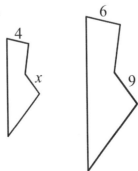

Chapter 18   Plane Geometry

18.13   Plane Geometry Word Problems

Solve the following word problems using the area, perimeter, and similar figure skills you have learned in this chapter.

1. Marvin has a piece of sheet metal measuring 3 feet by 2 feet. Marvin needs to remove a section on one end measuring $\frac{3}{4}$ of a foot by 2 feet. How much will remain of the sheet of metal?

2. Skeeter is helping with the scenery for the school play. He needs to paint a building on a sheet of plywood and show some sky at the top of the plywood. The sheet of plywood measures 4 feet by 8 feet. The sky painted at the top measure 4 feet by 1 foot. How much of the plywood is painted as a building?

3. Jose's dad is laying cement in a seating area of the garden. His dad wants to leave a flower bed area in the middle measuring 2 feet by 8 feet. The whole area measures 225 square feet. How many square feet of surface area will Jose's dad lay cement?

4. Mary is cutting out fabric to make a wall hanging for her room. The fabric measures 4 feet by 6 feet. She only needs 2 feet by 3 feet. How many square feet of fabric will Mary have leftover?

5. Karissa is drawing a garden design for her backyard. She has a plot 20 feet by 30 feet she will divide into 3 parts. The first part is a 15 feet by 15 feet, the second is a rectangle with an area of 175 ft^2. What is the area of the 3rd part?

6. Jeremy has cut a triangle with a base of 3 feet and a height of $1\frac{1}{2}$ feet out of a 4 foot square of plywood. Jeremy is planning on painting a design on the triangle and then hanging it in the basement to decorate the family room. How many square feet of plywood will Jeremy have left over?

7. Mr. Landish is trying to figure out the square footage of his lawn so he can buy the appropriate amount of fertilizer. He knows his lot is 110 feet by 220 feet, his house is 1,600 ft^2, and his driveway is 550 ft^2. What is the area of Mr. Landish's lawn?

8. Betty and Jake are figuring the amount of carpet needed to cover their daughter's bedroom. The three girls share a room measuring 13 feet by 12 feet, and there are two closets they want to carpet measuring 3 feet by 6 feet and the other is 3 feet by 3 feet. What is the total number of square feet of carpet needed to cover the floor and both closets?

9. The perimeter of a cage for the opossum exhibit at a local zoo is 60 feet. The length of the cage is 20 feet. What is the area of the base of the cage?

10. Ching Ngo is making a pair of drapes for her living room. Each of the four drapery panels, when completed, measures 8 feet long by 4 feet wide. What is the total area of the four drapery panels?

11. A painting includes an equilateral triangle measuring 8 cm on all sides. Next to it is a circle with a diameter measuring 8 cm. Are the two figures similar?

12. A rectangle measures 12 cm by 8 cm. A similar rectangle measures 9 cm on its longer side. What is the perimeter of the smaller rectangle?

13. A round table has a diameter of 60 inches. If Rhonda wants to make a round doily for the center of the table that is 20% of the size of the table, what will be the diameter of the doily?

222   Copyright © American Book Company

18.14 Perimeter and Area with Algebraic Expressions

You have already calculated the perimeter and area of various shapes with given measurements. You must also understand how to find the perimeter of shapes that are described by algebraic expressions. Study the examples below.

Example 10: Use the equation $P = 2l + 2w$ to find the perimeter of the following rectangle.

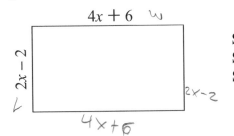

Step 1: Find $2l$. $2(4x + 6) = 8x + 12$
Step 2: Find $2w$. $2(2x - 2) = 4x - 4$
Step 3: Find $2l + 2w$. $12x + 8$

Perimeter $= 12x + 8$

Example 11: Using the formula $A = lw$, find the area of the rectangle below.

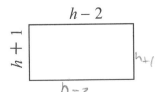

Step 1: $A = (h - 2)(h + 1)$
Step 2: $A = h^2 - 2h + h - 2$
Step 3: $A = h^2 - h - 2$

Area $= h^2 - h - 2$

Example 12: Find the area of a circle with $r = 4x + 1$.

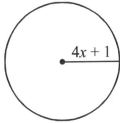

Step 1: Substitute the radius ($r = 4x + 1$) into the formula for a circle $A = \pi r^2$.
$A = \pi r^2 = \pi (4x + 1)^2$

Step 2: Simplify.
$A = \pi (4x + 1)^2 = \pi (4x + 1)(4x + 1) = \pi (16x^2 + 8x + 1)$
$A = \pi (16x^2 + 8x + 1)$

Chapter 18   Plane Geometry

Find the perimeter of each of the following figures.

1.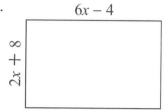

2.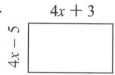

3.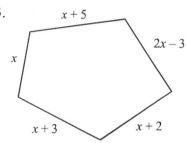

4. (4x + 3, 3x + 2)

5.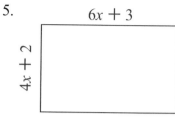

6.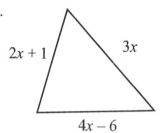

Find the area of each of the following rectangles.

7. (5 − 2m, 2 + m)

8.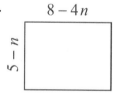

9. (2h − 2, 2h − 2)

10. (9 − h, 4 + 2h)

11. (n, n + 8)

12. (7 + 2b, 2 + b)

Find the area of the figures below.

13.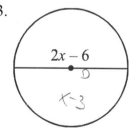

14.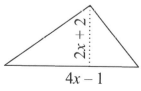

15.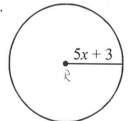

16.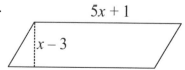

17.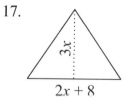

18.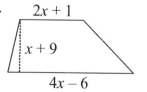

Chapter 18 Review

1. Calculate the perimeter of the following figure.

   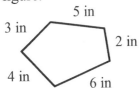

2. Find the area of the shaded region of the figure below.

   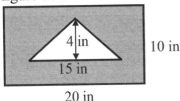

3. Calculate the perimeter and area.

   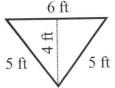

4. Calculate the perimeter and area.

   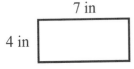

5. Find the area.

   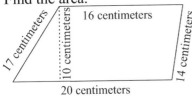

6. Find the area.

   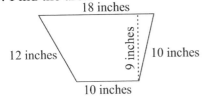

7. Find the area of the parallelogram.

   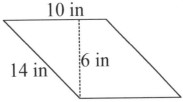

8. Find the missing side below.

   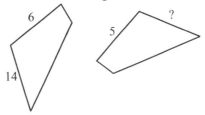

9. Find the area of the shaded part of the image below. Use $\pi \approx 3.14$.

   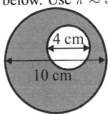

10. Calculate the circumference and the area of a circle with a radius = 7 cm. Use $\pi \approx \frac{22}{7}$.

11. Calculate the circumference and the area of a circle with a diameter = 2 ft. Use $\pi \approx 3.14$.

12. True or false: The opposite sides of a parallelogram are not congruent.

13. True or false: Two angles that correspond to the same arc are equal.

Chapter 18 Test

1. Find the area.

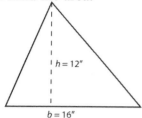

   A. 96 in
   B. 28 in^2
   C. 96 in^2
   D. 28 in
   E. 192 in^2

2. The figure below is a circle inscribed in a square. What is the area of the shaded region?

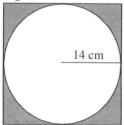

   F. 44 square centimeters
   G. 168 square centimeters
   H. 196 square centimeters
   J. 616 square centimeters
   K. 784 square centimeters

3. What is the area of the figure below?

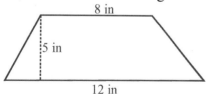

   A. 25 in^2
   B. 30 in^2
   C. 100 in^2
   D. 480 in^2
   E. 50 in^2

4. Using the formula $A = \frac{1}{2}bh$ for the area of a triangle, find the area of the triangle below.

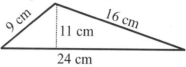

   F. 49 cm^2
   G. 132 cm^2
   H. 264 cm^2
   J. 3,456 cm^2
   K. 4,328 cm^2

5. What is the area of a square that measures 12 inches on each side?

   A. 48 inches^2
   B. 24 inches^2
   C. 132 inches^2
   D. 144 inches^2
   E. 148 inches^2

6. Find the area.

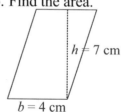

   F. 24 cm
   G. 24 cm^2
   H. 22 cm
   J. 22 cm^2
   K. 28 cm^2

7. How many square feet of sod are needed to cover a 9-foot by 60-foot lawn?

   A. 69 square feet
   B. 138 square feet
   C. 270 square feet
   D. 540 square feet
   E. 5,400 square feet

Chapter 18 Test

8. Find the perimeter.

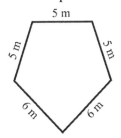

- F. 30 m
- G. 27 m
- H. 25 m
- J. 28 m
- K. 22 m

9. Find the perimeter.

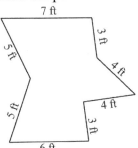

- A. 40 ft
- B. 38 ft
- C. 37 ft
- D. 36 ft
- E. 34 ft

10. Find the area of the shaded region.

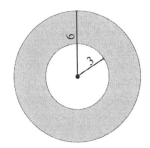

- F. 24π
- G. 3π
- H. 27π
- J. 15π
- K. 18π

11. Find the area of the shaded region.

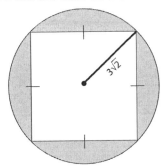

- A. $3\sqrt{2}\pi - 36$
- B. $18\pi - 36$
- C. $18\pi - 18$
- D. $3\sqrt{2}\pi - 18$
- E. $3\sqrt{2}\pi - 3$

12. What is the area of a circle with a radius of 7 cm? (Round to the nearest whole number)

- F. 154 square cm
- G. 196 square cm
- H. 347 square cm
- J. 616 square cm
- K. 49 square cm

13. Which of the following figures is a parallelogram?

A.

B.

C.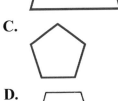

D.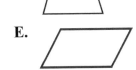

E.

Copyright © American Book Company

Chapter 18 Plane Geometry

14. Find the circumference of a circle with a radius that equals 5 cm. Use $\pi \approx 3.14$.

   F. 15.7 cm
   G. 62.8 cm
   H. 31.4 cm
   J. 0.314 cm
   K. 20.5 cm

15. Find the area. Use $\pi \approx 3.14$.

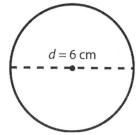

   A. 113.04 cm^2
   B. 28.26 cm^2
   C. 18.84 cm^2
   D. 188.4 cm^2
   E. 36 cm^2

16. Which item below is not a polygon?

   F. triangle
   G. heptagon
   H. octagon
   J. circle
   K. square

17. What is the name of the polygon below?

   A. quadrilateral 4
   B. pentagon 5
   C. hexagon 6
   D. octagon
   E. heptagon

18. What is the circumference of a circle that has a diameter of 10 cm?

   F. 15.7 cm
   G. 31.4 cm
   H. 78.5 cm
   J. 310 cm
   K. 100 cm

19. What is the measure of x?

   A. 10
   B. 13.5
   C. 26
   D. 6
   E. Cannot be determined without the measure of the other sides.

20. The length of the rectangle is 5 units longer than the width. Which expression could be used to represent the area of the rectangle?

   F. $w^2 + 5w$
   G. $w^2 + 5$
   H. $w^2 + 25$
   J. $w^2 + 10w + 25$
   K. $5w \times w$

21. Which of the following is the definition of a ray?

   A. A ray is a line segment with two endpoints.
   B. A ray goes on to infinity in both directions.
   C. A ray has one endpoint and goes to infinity in one direction.
   D. A ray is made up of one point.
   E. A ray always runs parallel to any line.

Chapter 19
Solid Geometry

19.1 Understanding Volume

Measurement of **volume** is expressed in cubic units such as in³, ft³, m³, cm³, or mm³. The volume of a solid is the number of cubic units that can be contained in the solid.

First, let's look at rectangular solids.

Example 1: How many 1 cubic centimeter cubes will it take to fill up the figure below?

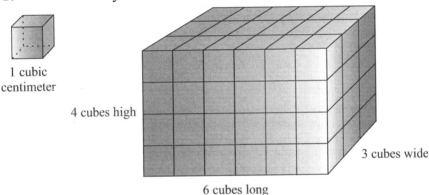

To find the volume, you need to multiply the length times the width times the height.
Volume of a right prism = area of the base × height ($V = Bh$), where $B = lw$.
$V = (6 \times 3) \times 4 = 72 \text{ cm}^3$ => B = (L×W)×h

19.2 Volume of Right Prisms

You can calculate the volume (V) of a right prism (box) by multiplying the area of the base ($B = lw$) by the height (h), as expressed in the formula $V = (Bh)$.

Example 2: Find the volume of the box pictured here:

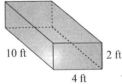

Step 1: Insert measurements from the figure into the formula.
Step 2: Multiply to solve. $(10 \times 4) \times 2 = 80 \text{ ft}^3$

Chapter 19   Solid Geometry

Find the volume of the following right prisms (boxes).

1.
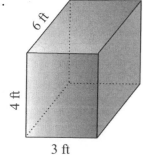

4.
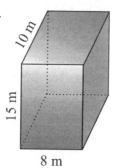

7.
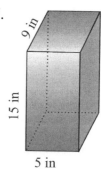

2.
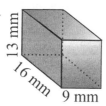

5.
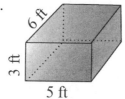

8.
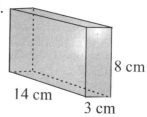

3.
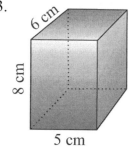

6.
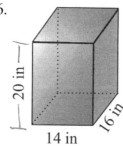

9.
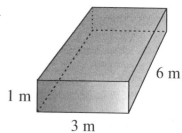

Find the missing information.

10.
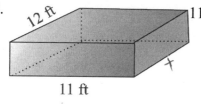

11.

12.
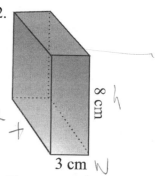

$V = 264$ ft^3
$h = $ _____

$V = 300$ m^3
$w = $ _____

$V = 192$ cm^3
$l = $ _____

230

19.3 Volume of Spheres, Cones, Cylinders, and Pyramids

To find the volume of a solid, insert the measurements given for the solid into the correct formula and solve. Remember, volumes are expressed in cubic units such as in^3, ft^3, m^3, cm^3, or mm^3.

Sphere
$$V = \frac{4}{3}\pi r^3$$

Cone
$$V = \frac{1}{3}\pi r^2 h$$

Cylinder
$$V = \pi r^2 h$$

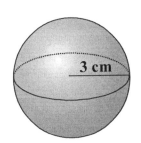
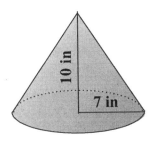
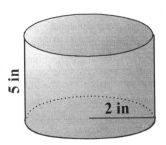

$V = \frac{4}{3}\pi r^3 \quad \pi \approx 3.14$
$V \approx \frac{4}{3} \times 3.14 \times 3^3$
$V \approx 113.04 \text{ cm}^3$

$V = \frac{1}{3}\pi r^2 h \quad \pi \approx 3.14$
$V \approx \frac{1}{3} \times 3.14 \times 7^2 \times 10$
$V \approx 512.87 \text{ in}^3$

$V = \pi r^2 h \quad \pi \approx \frac{22}{7}$
$V \approx \frac{22}{7} \times 2^2 \times 5$
$V \approx 62\frac{6}{7} \text{ in}^3$

Pyramids

$V = \frac{1}{3}Bh \quad B = $ area of rectangular base

$V = \frac{1}{3}Bh \quad B = $ area of triangular base

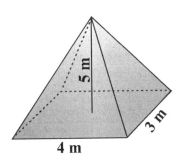
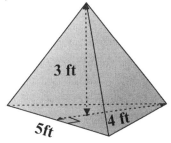

$V = \frac{1}{3}Bh \quad B = l \times w$
$V = \frac{1}{3} \times (4 \times 3) \times 5$
$V = 20 \text{ m}^3$

$V = \frac{1}{3}Bh \quad B = \frac{1}{2} \times b \times h$
$B = \frac{1}{2} \times 5 \times 4 = 10 \text{ ft}^2$
$V = \frac{1}{3} \times 10 \times 3$
$V = 10 \text{ ft}^3$

Copyright © American Book Company

Chapter 19   Solid Geometry

Find the volume of the following shapes. Use $\pi \approx 3.14$.

1.
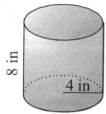
8 in, 4 in

6.
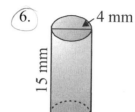
4 mm, 15 mm

2.
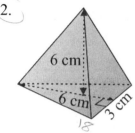
6 cm, 6 cm, 3 cm

7.
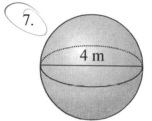
4 m

3.
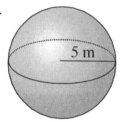
5 m

8.
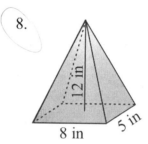
12 in, 8 in, 5 in

4.

8 ft, 2 ft

9.
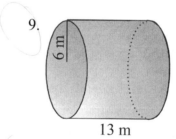
6 m, 13 m

5.
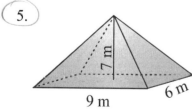
7 m, 9 m, 6 m

10.
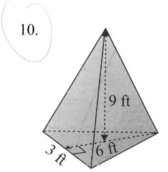
9 ft, 3 ft, 6 ft

Find x. **Use** $\pi \approx 3.14$.

11.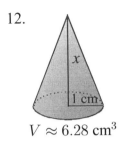

$V \approx 649.98 \text{ in}^3$

12.

$V \approx 6.28 \text{ cm}^3$

13.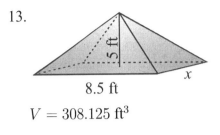

$V = 308.125 \text{ ft}^3$

19.4 Two-Step Volume Problems

Some objects are made from two geometric figures. For example, the tower below is made up of two geometric objects, a right prism and a pyramid.

Example 3: Find the volume of the tower.

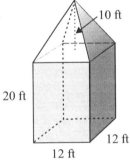

Step 1: Determine which formulas you will need. The tower is made from a pyramid and a right prism, so you will need the formulas for the volume of these two figures.

Step 2: Find the volume of each part of the tower. The bottom of the tower is a right prism $V = Bh$

$V = (12 \times 12) \times 20 = 2,880 \text{ ft}^3$

The top of the tower is a rectangular pyramid. $V = \frac{1}{3}Bh$

$V = \frac{1}{3} \times 12 \times 12 \times 10 = 480 \text{ ft}^3$

Step 3: Add the two volumes together. $2880 \text{ ft}^3 + 480 \text{ ft}^3 = 3,360 \text{ ft}^3$

Copyright © American Book Company

Find the volume of the geometric figures below. Hint: If part of a solid has been removed, find the volume of the hole, and subtract it from the volume of the total object.

1.

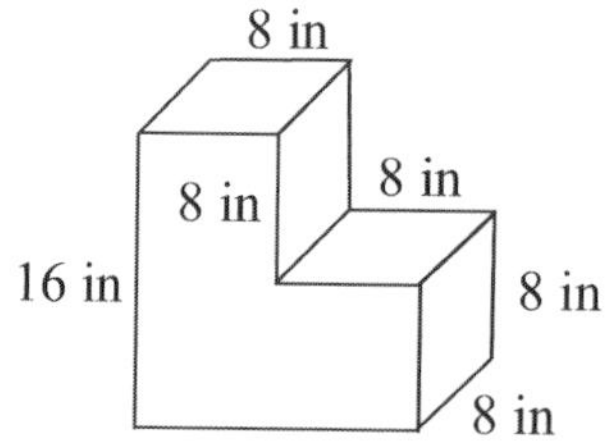

2. Each side measures three inches.

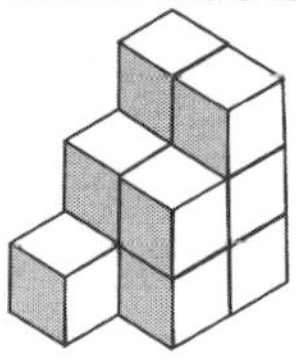

3. A rectangular hole passes through the middle of the figure below. The hole measures one cm on each side.

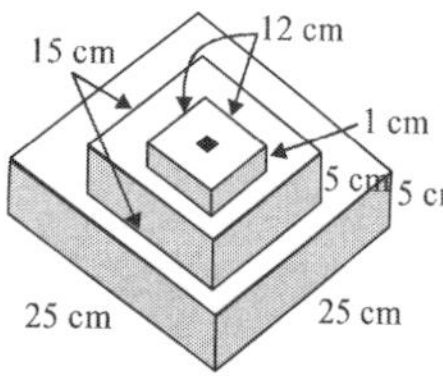

4. In the figure below, 3 cylinders are stacked on top of each other. The radii of the cylinders are 2 inches, 4 inches, and 6 inches. The height of each cylinder is one inch.

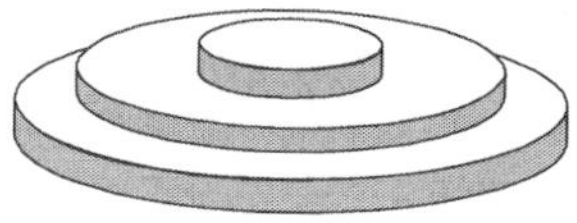

5.

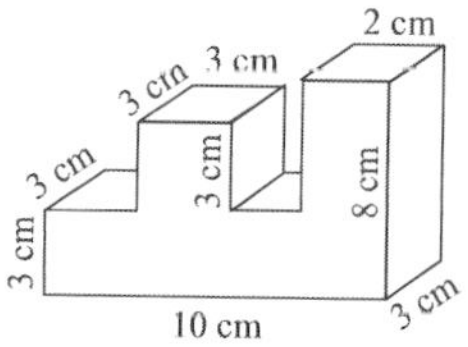

6. A hole, one meter in diameter, has been cut through the cylinder below.

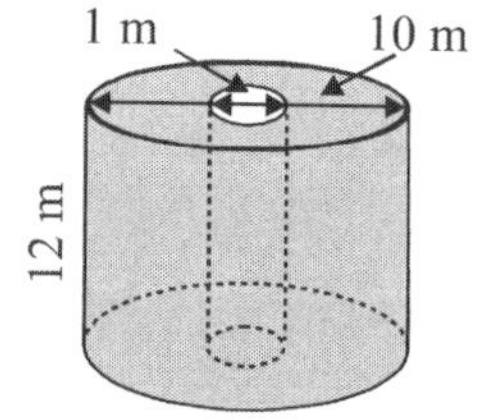

Find x. Use $\pi \approx 3.14$.

7. $V \approx 508.68$ in^3

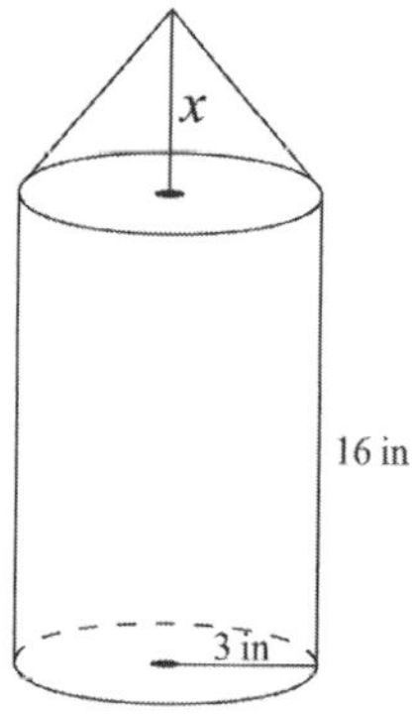

8. Volume of the shaded area ≈ 392.5 m^3

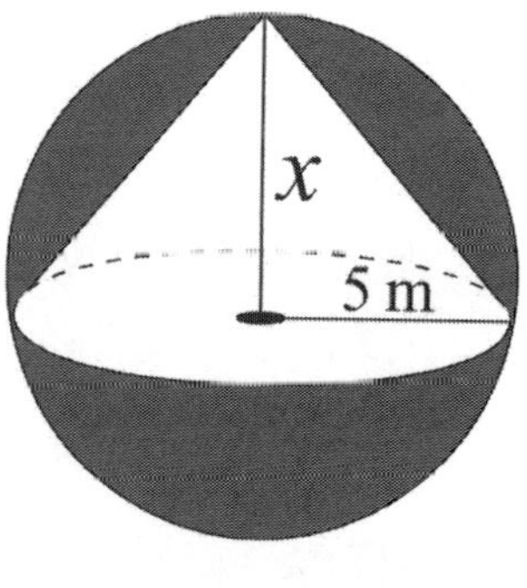

19.5 Solid Geometry Word Problems

1. If an Egyptian pyramid has a square base that measures 500 yards by 500 yards, and the pyramid stands 300 yards tall, what would be the volume of the pyramid? Use the formula for volume of a pyramid, $V = \frac{1}{3}Bh$ where B is the area of the base.

2. Robert is using a cylindrical barrel filled with water to flatten the sod in his yard. The circular ends have a radius of 1 foot. The barrel is 3 feet wide. How much water will the barrel hold? The formula for volume of a cylinder is $V = \pi r^2 h$. Use $\pi \approx 3.14$.

3. If a basketball measures 24 centimeters in diameter, what volume of air will it hold? The formula for volume of a sphere is $V = \frac{4}{3}\pi r^3$. Use $\pi \approx 3.14$.

4. What is the volume of a cone that is 2 inches in diameter and 5 inches tall? The formula for volume of a cone is $V = \frac{1}{3}\pi r^2 h$. Use $\pi \approx 3.14$.

5. Kelly has a rectangular fish aquarium that measures 24 inches wide, 12 inches deep, and 18 inches tall. What is the maximum amount of water that the aquarium will hold?

6. Jenny has a rectangular box that she wants to fill with beads. The box is 10 cm long, 5 cm wide, and 5 cm high. What is the volume of the box?

7. Gasco needs to construct a cylindrical, steel gas tank that measures 6 feet in diameter and is 8 feet long. What is the volume of the gas tank?

8. Craig wants to build a miniature replica of San Francisco's Transamerica Pyramid out of glass. His replica will have a square base that measures 6 cm by 6 cm. The 4 triangular sides will be 6 cm wide and 60 cm tall. What is the volume of Craig's pyramid?

9. Jeff built a wooden, cubic toy box for his son. Each side of the box measures 2 feet. How many cubic feet of toys will the box hold?

Copyright © American Book Company

235

Chapter 19 Solid Geometry

Chapter 19 Review

Find the volume of the following objects. If the volume is given in a problem, solve for x.

1.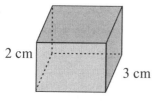
   2 cm, 3 cm, 3 cm
   $V =$ _____

2.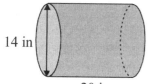
   14 in, 20 in
   Use $\pi \approx \frac{22}{7}$.
   $V \approx$ _____

3.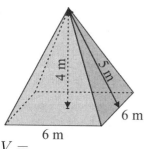
   4 m, 5 m, 6 m, 6 m
   $V =$ _____

4.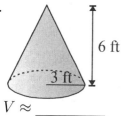
   6 ft, 3 ft
   $V \approx$ _____

5.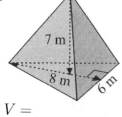
   7 m, 8 m, 6 m
   $V =$ _____

6. 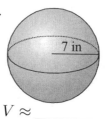 Use $\pi \approx \frac{22}{7}$.
   7 in
   $V \approx$ _____

7.
   x, 21 cm
   $V \approx 3231.06$ cm^3
   Find x.

8.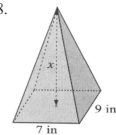
   x, 9 in, 7 in
   $V = 273$ in^3
   Find x.

9.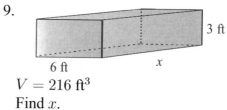
   3 ft, 6 ft, x
   $V = 216$ ft^3
   Find x.

10.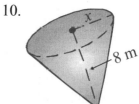
    x, 8 m
    $V = 6\pi$ m^3
    Find x.

Chapter 19 Review

11. The sandbox at the local elementary school is 60 inches wide and 100 inches long. The sand in the box is 6 inches deep. How many cubic inches of sand are in the sandbox?

12. If you have cubes that are two inches on each edge, how many would fit in a cube that was 16 inches on each edge?

13. If a ball is 4 inches in diameter, what is its approximate volume? Use $\pi \approx 3.14$.

14. A grain silo is in the shape of a cylinder. If the silo has an inside diameter of 10 feet and a height of 35 feet, what is the maximum volume inside the silo? Use $\pi \approx \frac{22}{7}$.

15. A closed cardboard box is 30 centimeters long, 10 centimeters wide, and 20 centimeters high. What is the volume of the box?

16. Siena wants to build a wooden toy box with a lid. The dimensions of the toy box are 3 feet long, 4 feet wide, and 2 feet tall. What is the volume of Siena's box?

17. How many 1-inch cubes will fit inside a larger 1 foot cube? (Figures are not drawn to scale.)

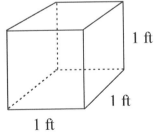

18. The cylinder below has a volume of 240 cubic inches. The cone below has the same radius and the same height as the cylinder. What is the volume of the cone?

19. Estimate the volume of the figure below.

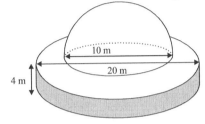

20. Find the volume of the figure below.

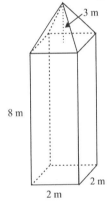

21. Find the volume of the figure below. Each side of each cube measures 4 feet.

22. A gigantic bronze sphere is being added to the top of a tall building downtown. The sphere will be 24 ft in diameter. What will be the approximate volume of the globe? Use $\pi \approx 3.14$.

Chapter 19   Solid Geometry

Chapter 19 Test

1. What is the volume, in cubic feet, of the square pyramid below?

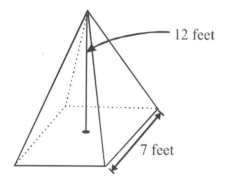

  A. 168 cubic feet
  B. 196 cubic feet
  C. 294 cubic feet
  D. 588 cubic feet
  E. 84 cubic feet

2. What is the approximate volume of the following oil tank? Round your answer to the nearest hundredth. Use $\pi \approx 3.14$.

  F. 18.84 yd^3
  G. 37.68 yd^3
  H. 44.48 yd^3
  J. 75.36 yd^3
  K. 12 yd^3

3. Find the approximate volume of the cone. Use the formula $V = \frac{1}{3}\pi r^2 h$, $\pi \approx \frac{22}{7}$.

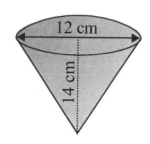

  A. 88 cm^3
  B. 176 cm^3
  C. 254 cm^3
  D. 528 cm^3
  E. 168 cm^3

4. What is volume of the box shown below?

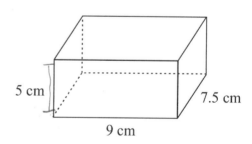

  F. 1875 cm^3
  G. 2250 cm^3
  H. 230 cm^3
  J. 345.5 cm^3
  K. 337.5 cm^3

Chapter 19 Test

5. If a hole with a 3 inch diameter is cut through a cylinder, what is the approximate volume afterwards?

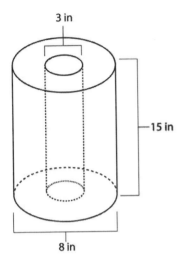

A. 375π in^3
B. 206.25π in^3
C. 240π in^3
D. 273.75π in^3
E. 360π in^3

6. If a sphere is cut out of a cube with a side measure of 12 m like the one shown below, about what would the new volume be?

 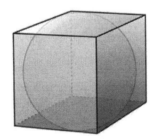

F. 823.22 m^3
G. 1728 m^3
H. 904.78 m^3
J. 1441 m^3
K. 216 m^3

7. The length of a cube's edge is 3 cm. Find the volume of the cube.

A. 36 cm^3
B. 9 cm^3
C. 18 cm^3
D. 12 cm^3
E. 27 cm^3

8. Find x.

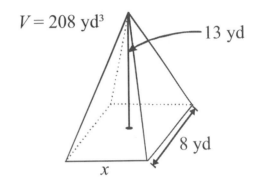

F. 8 yd
G. 16 yd
H. 6 yd
J. 48 yd
K. 32 yd

9. Find x.

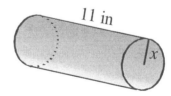

$V = 863.5$ in^3, $\pi \approx 3.14$

A. 15 in
B. 4 in
C. 5 in
D. 7 in
E. 34.255 in

Chapter 20
Transformations

20.1 Reflections

A **reflection** of a geometric figure is a mirror image of the object. Placing a mirror on the **line of reflection** will give you the position of the reflected image.

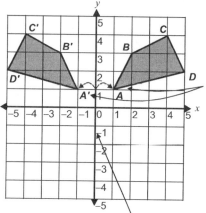

Quadrilateral *ABCD* is reflected across the y-axis to form quadrilateral $A'B'C'D'$. The y-axis is the line of reflection. Point A' (read as A prime) is the reflection of point A, point B' corresponds to point B, C' to C, and D' to D.

Point A is $+1$ space from the y-axis. Point A's mirror image, point A', is -1 space from the y-axis.

Point B is $+2$ spaces from the y-axis. Point B' is -2 spaces from the y-axis.

Point C is $+4$ spaces from the y-axis and point C' is -4 spaces from the y-axis.

Point D is $+5$ spaces from the y-axis and point D' is -5 spaces from the y-axis.

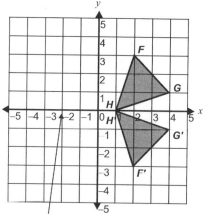

20.1 Reflections

Triangle *FGH* is reflected across the *x*-axis to form triangle *F'G'H'*. The *x*-axis is the line of reflection. Point *F'* reflects point *F*. Point *G'* corresponds to point *G*, and *H'* mirrors *H*. Point *F* is +3 spaces from the *x*-axis. Likewise, point *F'* is −3 spaces from the *x*-axis. Point *G* is +1 space from the *x*-axis, and point *G'* is −1 space from the *x*-axis. Point *H* is 0 spaces from the *x*-axis, so point *H'* is also 0 spaces from the *x*-axis.

Reflecting Across a 45° Line

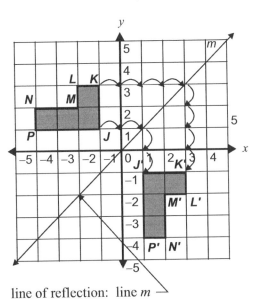

line of reflection: line *m*

Figure *JKLMNP* is reflected across line *m* to form figure *J'K'L'M'N'P'*. Line *m* is at a 45° angle. Point *J* corresponds to *J'*, *K* to *K'*, *L* to *L'*, *M* to *M'*, *N* to *N'* and *P* to *P'*. Line *m* is the line of reflection. **Pay close attention to how to determine the mirror image of figure *JKLMNP* across line *m* described below. This method only works when the line of reflection is at a 45° angle.**

Point *J* is 2 spaces over from line *m*, so *J'* must be 2 spaces down from line *m*.

Point *K* is 4 spaces over from line *m*, so *K'* is 4 spaces down from line *m*, and so on.

Draw the following reflections and record the new coordinates of the reflection. The first problem is done for you.

1. Reflect figure *ABC* across the *x*-axis. Label vertices *A'B'C'* so that point *A'* is the reflection of point *A*, *B'* is the reflection of *B*, and *C'* is the reflection of *C*.
   $A' = \underline{(-4, -2)}$ $B' = \underline{(-2, -4)}$ $C' = \underline{(0, -4)}$

2. Reflect figure *ABC* across the *y*-axis. Label vertices *A''B''C''* so that point *A''* is the reflection of point *A*, *B''* is the reflection of *B*, and *C''* is the reflection of *C*.
   $A'' = \underline{}$ $B'' = \underline{}$ $C'' = \underline{}$

3. Reflect figure *ABC* across line *p*. Label vertices *A'''B'''C'''* so that point *A'''* is the reflection of point *A*, *B'''* is the reflection of *B*, and *C'''* is the reflection of *C*.
   $A''' = \underline{}$ $B''' = \underline{}$ $C''' = \underline{}$

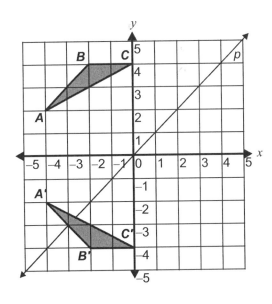

Copyright © American Book Company

241

Chapter 20 Transformations

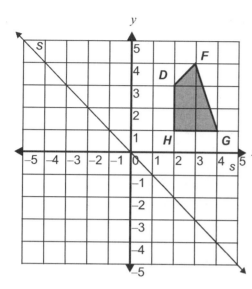

4. Reflect figure DFGH across the y-axis. Label vertices D'F'G'H' so that point D' is the reflection of point D, F' is the reflection of F, G' is the reflection of G, and H' is the reflection of H.
   D' = _____   G' = _____
   F' = _____   H' = _____

5. Reflect figure DFGH across the x-axis. Label vertices D'', F'', G'', H'' so that point D'' is the reflection of D, F'' is the reflection of F, G'' is the reflection of G, and H'' is the reflection of H.
   D'' = _____   G'' = _____
   F'' = _____   H'' = _____

6. Reflect figure DFGH across line s. Label vertices D'''F'''G'''H''' so that point D''' is the reflection of D, F''' corresponds to F, G''' to G, and H''' to H.
   D''' = _____   G''' = _____
   F''' = _____   H''' = _____

7. Reflect quadrilateral MNOP across the y-axis. Label vertices M'N'O'P' so that point M' is the reflection of point M, N' is the reflection of N, O' is the reflection of O, and P' is the reflection of P.
   M' = _____   O' = _____
   N' = _____   P' = _____

8. Reflect figure MNOP across the x-axis. Label vertices M'', N'', O'', P'' so that point M'' is the reflection of M, N'' is the reflection of N, O'' is the reflection of O, and P'' is the reflection of P.
   M'' = _____   O'' = _____
   N'' = _____   P'' = _____

9. Reflect figure MNOP across line w. Label vertices M'''N'''O'''P''' so that point M''' is the reflection of M, N''' corresponds to N, O''' to O, and P'''' to P.
   M''' = _____   O''' = _____
   N''' = _____   P''' = _____

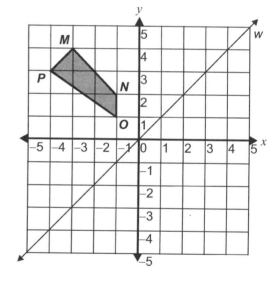

20.2 Translations

To make a translation of a geometric figure, first duplicate the figure and then slide it along a path.

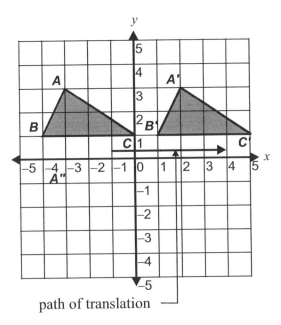

path of translation

Triangle $A'B'C'$ is a translation of triangle ABC. Each point is translated 5 spaces to the right. In other words, the triangle slid 5 spaces to the right. Look at the path of translation. It gives the same information as above. Count the number of spaces across given by the path of translation, and you will see it represents a move 5 spaces to the right. Each new point is found at $(x+5, y)$.

Point A is at $(-3, 3)$. Therefore, A' is found at $(-3+5, 3)$ or $(2, 3)$.

B is at $(-4, 1)$, so B' is at $(-4+5, 1)$ or $(1, 1)$.

C is at $(0, 1)$, so C' is at $(0+5, 1)$ or $(5, 1)$.

Quadrilateral $FGHI$ is translated 5 spaces to the right and 3 spaces down. The path of translation shows the same information. It points right 5 spaces and down 3 spaces. Each new point is found at $(x+5, y-3)$.

Point F is located at $(-4, 3)$. Point F' is located at $(-4+5, 3-3)$ or $(1, 0)$.

Point G is at $(-2, 5)$. Point G' is at $(-2+5, 5-3)$ or $(3, 2)$.

Point H is at $(-1, 4)$. Point H' is at $(-1+5, 4-3)$ or $(4, 1)$.

Point I is at $(-1, 2)$. Point I' is at $(-1+5, 2-3)$ or $(4, -1)$.

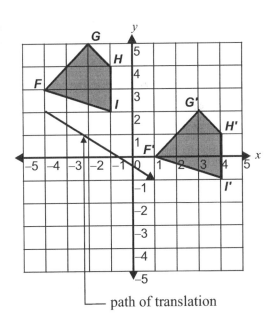

path of translation

Copyright © American Book Company

243

Chapter 20 Transformations

Draw the following translations and record the new coordinates of the translation. The figure for the first problem is drawn for you.

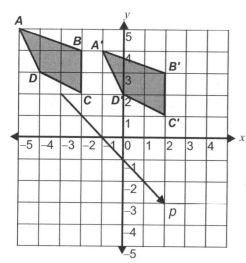

1. Translate figure ABCD 4 spaces to the right and 1 space down. Label the vertices of the translated figure A', B', C', and D' so that point A' corresponds to the translation of point A, B' corresponds to B, C' to C, and D' to D.

    $A' = $ _____ $C' = $ _____
    $B' = $ _____ $D' = $ _____

2. Translate figure ABCD 5 spaces down. Label the vertices of the translated figure A'', B'', C'', and D'' so that point A'' corresponds to the translation of point A, B'' corresponds to B, C'' to C, and D'' to D.

    $A'' = $ _____ $C'' = $ _____
    $B'' = $ _____ $D'' = $ _____

3. Translate figure ABCD along the path of translation, p. Label the vertices of the translated figure A''', B''', C''', and D''' so that point A''' corresponds to the translation of point A, B''' corresponds to B, C''' to C, and D''' to D.

    $A''' = $ _____ $C''' = $ _____
    $B''' = $ _____ $D''' = $ _____

4. Translate triangle FGH 6 spaces to the left and 3 spaces up. Label the vertices of the translated figure F', G', and H' so that point F' corresponds to the translation of point F, G' corresponds to G, and H' to H.

    $F' = $ _____ $G' = $ _____ $H' = $ _____

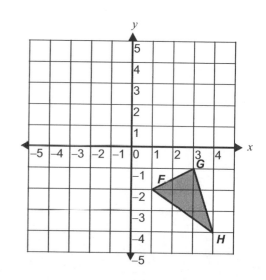

5. Translate triangle FGH 4 spaces up and 1 space to the left. Label the vertices of the translated triangle $F''G''H''$ so that point F'' corresponds to the translation of point F, G'' corresponds to G, and H'' to H.

    $F'' = $ _____ $G'' = $ _____ $H'' = $ _____

244   Copyright © American Book Company

20.3 Rotations

A **rotation** of a geometric figure shows motion around a point.

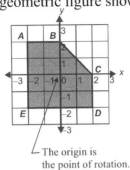

- The origin is the point of rotation.
- Figure *ABCDE* has been rotated $\frac{1}{4}$ of a turn clockwise around the origin to form *A'B'C'D'E'*.
- Figure *ABCDE* has been rotated $\frac{1}{2}$ of a turn around the origin to form *A''B''C''D''E''*.

Draw the following rotations, and record the new coordinates of the rotation. The figure for the first problem is drawn for you.

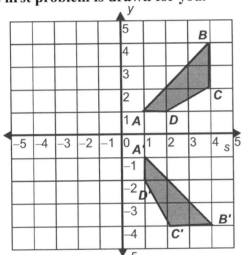

1. Rotate figure *ABCD* around the origin clockwise $\frac{1}{4}$ turn. Label the vertices *A'*, *B'*, *C'*, and *D'* so that point *A'* corresponds to the rotation of point *A*, *B'* corresponds to *B*, *C'* to *C*, and *D'* to *D*.
   A' = _____   C' = _____
   B' = _____   D' = _____

2. Rotate figure *ABCD* around the origin clockwise $\frac{1}{2}$ turn. Label the vertices *A''*, *B''*, *C''*, and *D''* so that point *A''* corresponds to the rotation of point *A*, *B''* corresponds to *B*, *C''* to *C*, and *D''* to *D*.
   A'' = _____   C'' = _____
   B'' = _____   D'' = _____

3. Rotate figure *ABCD* around the origin clockwise $\frac{3}{4}$ turn. Label the vertices *A'''*, *B'''*, *C'''*, and *D'''* so that point *A'''* corresponds to the rotation of point *A*, *B'''* corresponds to *B*, *C'''* to *C*, and *D'''* to *D*.
   A''' = _____   C''' = _____
   B''' = _____   D''' = _____

4. Rotate figure *MNO* around point *O* clockwise $\frac{1}{4}$ turn. Label the vertices *M'*, *N'*, and *O* so that point *M'* corresponds to the rotation of point *M* and *N'* corresponds to *N*.
   M' = _____   N' = _____

5. Rotate figure *MNO* around point *O* clockwise $\frac{1}{2}$ turn. Label the vertices *M''*, *N''*, and *O* so that point *M''* corresponds to the rotation of point *M*, and *N''* corresponds to *N*.
   M'' = _____   N'' = _____

6. Rotate figure *MNO* around point *O* clockwise $\frac{3}{4}$ turn. Label the vertices *M'''*, *N'''*, and *O* so that point *M'''* corresponds to the rotation of point *M*, and *N'''* corresponds to *N*.
   M''' = _____   N''' = _____

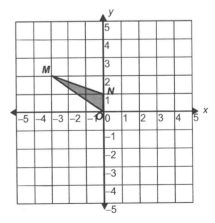

Chapter 20   Transformations

20.4   Transformation Practice

Answer the following questions regarding transformations.

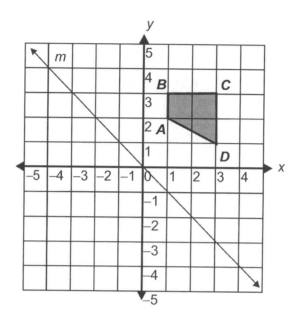

1. Translate quadrilateral *ABCD* so that point *A′*, which corresponds to point *A*, is located at coordinates $(-4, 3)$. Label the other vertices *B′* to correspond to *B*, *C′* to *C*, and *D′* to *D*. What are the coordinates of *B′*, *C′*, and *D′* ?

   $A' = $ _____ $C' = $ _____

   $B' = $ _____ $D' = $ _____

2. Reflect quadrilateral *ABCD* across line *m*. Label the coordinates *A″*, *B″*, *C″*, and *D″*, so that point *A″* corresponds to the reflection of point *A*, *B″* corresponds to the reflection of *B*, and *C″* corresponds to the reflection of *C*. What are the coordinates of *A″*, *B″*, *C″*, and *D″* ?

   $A'' = $ _____ $C'' = $ _____

   $B'' = $ _____ $D'' = $ _____

3. Rotate quadrilateral *ABCD* $\frac{1}{4}$ turn counterclockwise around point *D*. Label the points *A‴B‴C‴D‴* so that *A‴* corresponds to the rotation of point *A*, *B‴* corresponds to *B*, *C‴* to *C*, and *D‴* to *D*. What are the coordinates of *A‴*, *B‴*, *C‴* and *D‴* ?

   $A''' = $ _____ $C''' = $ _____

   $B''' = $ _____ $D''' = $ _____

246   Copyright © American Book Company

20.5 Dilations

A **dilation** of a geometric figure is either an enlargement or a reduction of the figure. The point at which the figure is either reduced or enlarged is called the center of dilation. The dilation of a figure is always the product of the original and a **scale factor**. The scale factor is always a positive number that is multiplied by the coordinates of a shape's vertices, which is usually illustrated in a coordinate plane. If the scale factor is greater than one, then the resulting dilated figure will be an enlargement of the original figure. If the scale factor is less than one, then the resulting dilated figure will be a reduction of the original figure.

Example 1: The triangle ABC has been dilated by a scale factor of $\frac{1}{4}$.

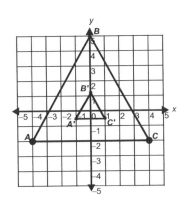

The first step in finding the dilated object is to list all the vertices of the original object, $\triangle ABC$. The next step is to multiply the coordinates of the vertices of $\triangle ABC$ by the scale factor, $\frac{1}{4}$, to find the coordinates of the dilated figure. Lastly, draw the dilated object on the coordinate plane as shown above.

$A : (-4, -2)$ $A' : \left(-1, -\frac{1}{2}\right)$
$B : (0, 5)$ $B' : \left(0, \frac{5}{4}\right)$
$C : (4, -2)$ $C' : \left(1, -\frac{1}{2}\right)$

Note: Since the scale factor is less than one, the dilated figure $A'B'C'$ is a reduction of original triangle, ABC.

Circle the coordinate plane that contains the shape that has been dilated.

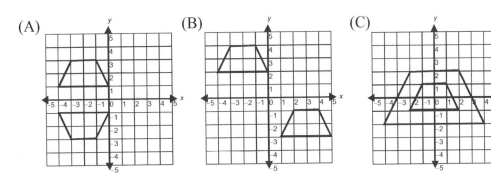

Chapter 20   Transformations

On your own graph paper sketch the dilated and original figures.

For the questions 1–6, find the coordinates of the vertices of the dilated figure.

1. A: $(-3, 1)$
   B: $(-1, 4)$
   C: $(1, 4)$
   D: $(3, 1)$
   Scale factor: 4

2. A: $(-6, 5)$
   B: $(3, 5)$
   C: $(3, -4)$
   D: $(-6, -4)$
   Scale factor: $\frac{1}{3}$

3. A: $(-10, 0)$
   B: $(0, 10)$
   C: $(8, 5)$
   Scale factor: $\frac{4}{5}$

4. A: $(-1, 7)$
   B: $(1, 7)$
   C: $(5, 5)$
   D: $\left(5, \frac{1}{2}\right)$
   E: $(1, -3)$
   F: $(-1, -3)$
   G: $\left(-5, \frac{1}{2}\right)$
   H: $(-5, 5)$
   Scale factor: 2

5. A: $(-8, 7)$
   B: $(-4, 7)$
   C: $(-2, 3)$
   D: $(-6, 3)$
   Scale factor: $\frac{3}{2}$

6. A: $(-4, 12)$
   B: $(6, -2)$
   C: $(-14, -2)$
   Scale factor: $\frac{1}{2}$

For questions 7–10, find the scale factor.

7. A: $(-3, 2)$ A': $(-10.5, 7)$
   B: $(1, 2)$ B': $(3.5, 7)$
   C: $(1, -3)$ C': $(3.5, -10.5)$
   D: $(-3, -3)$ D': $(-10.5, -10.5)$

8. A: $(-6, 9)$ A': $(-2, 3)$
   B: $(3, 12)$ B': $(1, 4)$
   C: $(6, 3)$ C': $(2, 1)$
   D: $(-9, 0)$ D': $(-3, 0)$

9. A: $(0, -3)$ A': $(0, -2)$
   B: $(6, 0)$ B': $(4, 0)$
   C: $(0, 3)$ C': $(0, 2)$

10. A: $(-2, 6)$ A': $(-10, 30)$
    B: $(2, 6)$ B': $(10, 30)$
    C: $(3, 3)$ C': $(15, 15)$
    D: $(2, 0)$ D': $(10, 0)$
    E: $(-2, 0)$ E': $(-10, 0)$
    F: $(-3, 3)$ F': $(-15, 15)$

For questions 11 and 12, determine whether or not $A'B'C'D'$ is a dilation of $ABCD$.

11. A: $(-2, 5)$ A': $(-1, 2)$
    B: $(8, 8)$ B': $(4, 4)$
    C: $(12, 0)$ C': $(6, 0)$
    D: $(2, -6)$ D': $(1, -3)$

12. A: $(0, 8)$ A': $(-2, 6)$
    B: $(5, 8)$ B': $(3, 6)$
    C: $(5, -3)$ C': $(3, -1)$
    D: $(0, -3)$ D': $(-2, -1)$

248   Copyright © American Book Company

Chapter 20 Review

1. Draw the reflection of image *ABCD* over the *y*-axis. Label the points *A′*, *B′*, *C′*, and *D′*. List the coordinates of these points below.

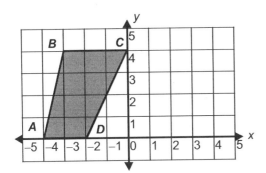

2. *A′* = _____
3. *B′* = _____
4. *C′* = _____
5. *D′* = _____

6. Rotate the figure below a $\frac{1}{2}$ turn about the origin. Label the points *A′*, *B′*, *C′*, *D′*, *E′*, and *F′*. List the coordinates of these points below.

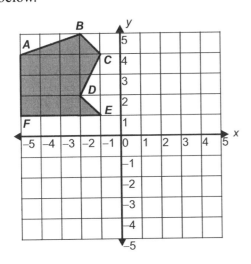

7. *A′* = _____
8. *B′* = _____
9. *C′* = _____
10. *D′* = _____
11. *E′* = _____
12. *F′* = _____

13. Use the translation described by the arrow to translate the polygon below. Label the points *P′*, *Q′*, *R′*, *S′*, *T′*, and *U′*. List the coordinate of each.

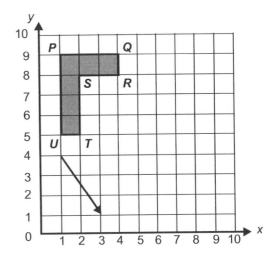

14. *P′* = _____
15. *Q′* = _____
16. *R′* = _____
17. *S′* = _____
18. *T′* = _____
19. *U′* = _____

Use the grid to answer the question that follows.

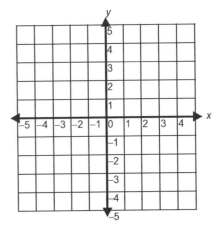

20. A point at $(-3, 2)$ is moved to $(0, 0)$. If a point at $(1, 1)$ is moved in the same way, what will its new coordinates be?

Chapter 20   Transformations

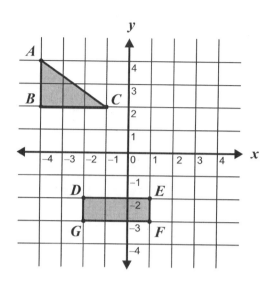

Find the coordinates of the geometric figures graphed above.

21. point A

22. point B

23. point C

24. point D

25. point E

26. point F

27. point G

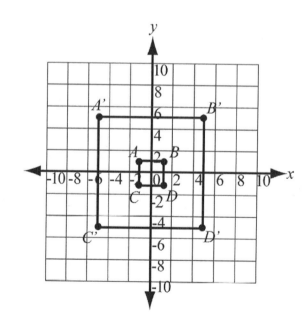

28. Are the figures plotted in the graph above an example of a reflection, rotation, transformation, or dilation?

250

Chapter 20 Test

1. If the figure below were reflected across the y-axis, what would be the coordinates of point C?

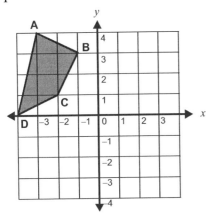

   A. $(2, -1)$
   B. $(1, 2)$
   C. $(-2, -1)$
   D. $(-2, 1)$
   E. $(2, 1)$

2. If the figure below is translated in the direction described by the arrow, what will be the new coordinates of point D after the transformation?

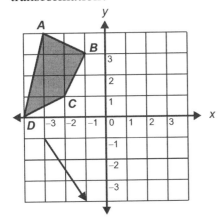

   F. $(-2, -4)$
   G. $(-1, -3)$
   H. $(-1, -4)$
   J. $(-2, -3)$
   K. $(-1, -2)$

3. Figure 1 goes through a transformation to form Figure 2. Which of the following descriptions fits the transformation(s) shown below?

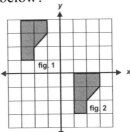

   A. reflection across the x-axis
   B. reflection across the origin
   C. $\frac{1}{2}$ clockwise rotation around the origin
   D. translation right 4 units and down 4 units
   E. rotation of $180°$

4. Sammy plots the point $(-4, 3)$ on a coordinate grid. He reflects this point over the y-axis, then over the x-axis. What are the coordinates of the new reflected point?

   F. $(-4, -3)$
   G. $(-4, 3)$
   H. $(4, -4)$
   J. $(4, 3)$
   K. $(4, -3)$

5. Figure 1 goes through a transformation to form figure 2. Which of the following descriptions fits the transformation shown?

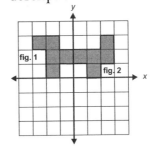

   A. reflection across the y-axis
   B. $\frac{3}{4}$ clockwise rotation around the origin
   C. translation right 3 units
   D. $\frac{1}{4}$ clockwise rotation around the origin
   E. $\frac{1}{2}$ clockwise rotation around the origin

Copyright © American Book Company

Chapter 21
Statistics

21.1   Range

In **statistics**, the difference between the largest number and the smallest number in a list is called the **range**. Range is a measure of variability. **Variability** is the measure of how disperse the data set is. If the data is spread out over a wide range of numbers, then the data has high variability. If the data is not spread out over a wide range of numbers, then the data has low variability.

Example 1:   Find the range of the following list of numbers: 16, 73, 26, 15, and 35.
The largest number is 73, and the smallest number is 15. $73 - 15 = 58$
The range is 58.

Find the range for each list of numbers below.

1.	21	2.	6	3.	89	4.	41	5.	23	6.	2	7.	77
	51		7		22		3		20		38		94
	48		31		65		56		64		29		27
	42		55		36		41		38		33		46
	12		8		20		19		21		59		63

8.	51	9.	65	10.	84	11.	84	12.	21	13.	45	14.	62
	62		54		59		65		78		57		39
	32		56		48		32		6		57		96
	16		5		21		50		97		14		45
	59		63		80		71		45		61		14

15. 2, 15, 3, 25, and 17

16. 15, 48, 52, 41, and 8

17. 54, 74, 2, 86, and 75

18. 15, 61, 11, 22, and 65

19. 33, 18, 65, 12, and 74

20. 47, 12, 33, 25, and 19

21. 56, 10, 33, 7, 16, and 5

22. 46, 25, 78, 49, and 6

23. 45, 75, 63, and 21

24. 97, 23, 56, 12, and 66

252   Copyright © American Book Company

21.3   Finding Data Missing From the Mean

21.2   Mean

In statistics, the mean is the same as the average. To find the mean of a list of numbers, first add together all of the numbers in the list, and then divide by the number of items in the list.

Example 2:   Find the mean of 38, 72, 110, 548.

   Step 1:   First add: $38 + 72 + 110 + 548 = 768$

   Step 2:   There are 4 numbers in the list so divide the total by 4. $768 \div 4 = 192$
                   The mean is 192.

Practice finding the mean. Round to the nearest tenth if necessary.

1. Dinners served:
    489 561 522 450

2. Prices paid for shirts:
    $4.89 $9.97 $5.90 $8.64

3. Student absences:
    6 5 13 8 9 12 7

4. Paychecks:
    $89.56 $99.99 $56.54

5. Long distance calls:
    33 14 24 21 19

6. Train boxcars:
    56 55 48 61 51

Find the mean of the following word problems.

7. Val's science grades are 95, 87, 65, 94, 78, and 97. What is her average?

8. Ann runs a business from her home. The number of orders for the last 7 business days are 17, 24, 13, 8, 11, 15, and 9. What is the average number of orders per day?

21.3   Finding Data Missing From the Mean

Example 3:   Mara knew she had an 88 average in her biology class, but she lost one of her papers. The three papers she could find had scores of 98%, 84%, and 90%. What was the score on her fourth paper?

   Step 1:   Calculate the total score on four papers with an 88% average. $.88 \times 4 = 3.52$

   Step 2:   Add together the scores from the three papers you have. $.98 + .84 + .9 = 2.72$

   Step 3:   Subtract the scores you know from the total score. $3.52 - 2.72 = 0.80$. She had 80% on her fourth paper.

Find the data missing from the following problems.

1. Gabriel earns 87% on his first geography test. He wants to earn a 92% average. What does he need to get on his next test to bring his average up?

2. Rian earned $68.00 on Monday. How much money must she earn on Tuesday to have an average of $80 earned for the two days?

3. Haley, Chuck, Dana, and Chris enter a contest to see who could bake the most chocolate chip cookies in an hour. They bake an average of 75 cookies. Haley bakes 55, Chuck bakes 70, and Dana bakes 90. How many does Chris bake?

Copyright © American Book Company

253

Chapter 21   Statistics

21.4   Median

In a list of numbers ordered from lowest to highest, the **median** is the middle number. To find the **median**, first arrange the numbers in numerical order. If there is an odd number of items in the list, the **median** is the middle number. If there is an even number of items in the list, the **median is the average of the two middle numbers.**

Example 4:   Find the median of 42, 35, 45, 37, and 41.

    Step 1:   Arrange the numbers in numerical order: 35 37 $\boxed{41}$ 42 45

    Step 2:   Find the middle number. The median is 41.

Example 5:   Find the median of 14, 53, 42, 6, 14, and 46.

    Step 1:   Arrange the numbers in numerical order: 6 14 $\boxed{14\ 42}$ 46 53.

    Step 2:   Find the average of the two middle numbers.
        $(14 + 42) \div 2 = 28$. The median is 28.

Find the median in each list of numbers.

1. 35, 55, 40, 30, and 45

2. 7, 2, 3, 6, 5, 1, and 8

3. 65, 42, 60, 46, and 90

4. 15, 16, 19, 25, and 20

5. 10, 8, 21, 14, 9, and 12

6. 43, 36, 20, and 40

7. 5, 24, 9, 18, 12, and 3

8. 48, 13, 54, 82, 90, and 7

9. 23, 21, 36, and 27

21.5   Mode

In statistics, the **mode is the number that occurs most frequently in a list of numbers**. If no number occurs more than once, there is no mode.

Example 6:   Exam grades for a math class were as follows:
    70 88 92 85 99 85 70 85 99 100 88 70 99 88 88 99 88 92 85 88

    Step 1:   Count the number of times each number occurs in the list.

        70 - 3 times
        88 - 6 times
        92 - 2 times
        85 - 4 times
        99 - 4 times
      100 - 1 time

    Step 2:   Find the number that occurs most often.
        The mode is 88 because it is listed 6 times. No other number is listed as often.

Find the mode in each of the following lists of numbers.

1. 48, 32, 56, 32, 56, 48, 56

2. 12, 16, 54, 78, 16, 25, 20

3. 5, 4, 8, 3, 4, 2, 7, 8, 4, 2

4. 11, 9, 7, 11, 7, 5, 7, 7, 5

5. 84, 22, 79, 22, 87, 22, 22

6. 95, 87, 65, 94, 78, 95

7. 8, 2, 5, 4, 7, 2, 3, 6, 1

8. 89, 7, 11, 89, 17, 56

9. 15, 48, 52, 41, 8, 48

254   Copyright © American Book Company

21.6   Applying Measures of Central Tendency

21.6   Applying Measures of Central Tendency

On the PLAN, you may be asked to solve real-world problems involving measures of central tendency.

Example 7:   Aida is shopping around for the best price on a 17" computer monitor. She travels to seven stores and finds the following prices: $199, $159, $249, $329, $199, $209, and $189. When Aida goes to the eighth and final store, she finds the price for the 17" monitor is $549. Which of the measures of central tendency, mean, median, or mode, changes the most as a result of the last price Aida finds?

Step 1:   **Solve for all three measures of the seven values.**

Mean: $\dfrac{\$199 + \$159 + \$249 + \$329 + \$199 + \$209 + \$189}{7} = \$219$

Median: From least to greatest: $159, $189, $199, $199, $209, $249, $329. The 4th value = $199

Mode: The number repeated the most is $199.

Step 2:   **Find the mean, median, and mode with the eighth value included.**

Mean: $\dfrac{\$199 + \$159 + \$249 + \$329 + \$199 + \$209 + \$189 + \$549}{8} = \$260.25$

Median: $159, $189, $199, $199, $209, $249, $329, $549. The avg. of 4th and 5th number = $204

Mode: The number still repeated most is $199.

Answer: The measure which changed the most by adding the 8th value is the **mean**.

1. The Realty Company has the selling prices for 10 houses sold during the month of July. The following prices are given in thousands of dollars:

   176   89   525   125   107   100   525   61   75   114

   Find the mean, median, and mode of the selling prices. Which measure is most representative for the selling price of such homes? Explain.

2. A soap manufacturing company wants to know if the weight of its product is on target, meaning 4.75 oz. With that purpose in mind, a quality control technician selects 15 bars of soap from production, 5 from each shift, and finds the following weights in oz.

   1st shift: 4.76, 4.75, 4.77, 4.77, 4.74
   2nd shift: 4.72, 4.72, 4.75, 4.76, 4.73
   3rd shift: 4.76, 4.76, 4.77, 4.76, 4.76

   (A) What are the values for the measures of central tendency for the sample from each shift?
   (B) Find the mean, median, and mode for the 24 hour production sample.
   (C) Which measure is the most accurate measure of central tendency for the 24 hour production?
   (D) Find the range of values for each shift. Is the range an effective tool for drawing a conclusion in this case? Why or why not?

Copyright © American Book Company

Chapter 21   Statistics

Chapter 21 Review

Find the mean, median, mode, and range for each of the following sets of data. Fill in the table below.

❶ Miles Run by Track Team Members	
Jeff	24
Eric	20
Craig	19
Simon	20
Elijah	25
Rich	19
Marcus	20

❷ 1992 SUMMER OLYMPIC GAMES Gold Medals Won			
Unified Team	45	Hungary	11
United States	37	South Korea	12
Germany	33	France	8
China	16	Australia	7
Cuba	14	Japan	3
Spain	13		

❸ Hardware Store Payroll June Week 2	
Erica	$280
Dane	$206
Sam	$240
Nancy	$404
Elsie	$210
Gail	$305
David	$280

Data Set Number	Mean	Median	Mode	Range
❶				
❷				
❸				

4. Jenica bowls three games and scores an average of 116 points per game. She scores 105 on her first game and 128 on her second game. What does she score on her third game?

5. Concession stand sales for each game in season are $320, $540, $230, $450, $280, and $580. What is the mean sales per game?

6. Cendrick D'Amitrano works Friday and Saturday delivering pizza. He delivers 8 pizzas on Friday. How many pizzas must he deliver on Saturday to average 11 pizzas per day?

7. Long cooks three Vietnamese dinners that weigh a total of 40 ounces. What is the average weight for each dinner?

8. The Swamp Foxes scored an average of 7 points per soccer game. They scored 9 points in the first game, 4 points in the second game, and 5 points in the third game. What was their score for their fourth game?

9. Shondra is 66 inches tall, and DeWayne is 72 inches tall. How tall is Michael if the average height of these three students is 67 inches?

Use the figure to help answer questions 10–13.

Jessica	92	85	97	87	88	86	91	93	89	89
Michael	91	89	83	95	80	91	81	96	93	93

10. What is the lowest score Jessica can receive on her next test in order to have a 90 test average?

11. What score does Michael need to make to get a 90% average?

12. Find the means.

13. Find the ranges.

256   Copyright © American Book Company

Chapter 21 Test

1. What is the mean of 36, 54, 66, 45, 36, 36, and 63?

   A. 36
   B. 45
   C. 48
   D. 63
   E. 66

2. A neighborhood surveyed the times of day people water their lawns and tallied the data below.

Time	Tally
midnight - 3:59 a.m.	II
4:00 a.m. - 7:59 a.m.	JHT I
8:00 a.m. - 11:59 a.m.	JHT IIII
noon - 3:59 p.m.	JHT
4:00 p.m. - 7:59 p.m.	JHT JHT
8:00 p.m. - 11:59 p.m.	JHT III

   If you wanted to find which was the most popular time of day to water the lawn, it would be best to find the _____ of data.

   F. mean
   G. median
   H. range
   J. mode
   K. None of the above.

3. Examine the following two data sets:

   Set #1: 49, 55, 68, 72, 98

   Set #2: 20, 36, 47, 68, 75, 82, 89

   Which of the following statements is true?

   A. They have the same mode.
   B. They have the same median.
   C. They have the same mean.
   D. They have the same range.
   E. None of the above.

4. What is the median of the following set of data?

   33, 31, 35, 24, 38, 30

   F. 38
   G. 31
   H. 30
   J. 29
   K. 32

5. Concession stand sales for the first 6 games of the season averaged $400.00. If the total sales of the first 5 games were $320, $540, $230, $450, and $280, what were the total sales for the sixth game?

   A. $230
   B. $350
   C. $364
   D. $580
   E. $400

6. Which of the following sets of numbers has a range of 51?

   F. {29, 19, 72, 68, 39}
   G. {81, 85, 37, 41, 60}
   H. {17, 12, 9, 47, 82}
   J. {62, 86, 44, 78, 95}
   K. {58, 86, 44, 78, 99}

7. What is the mean of 12, 23, 8, 26, 37, 11, and 9?

   A. 12
   B. 29
   C. 18
   D. 19
   E. 37

Copyright © American Book Company

Chapter 21   Statistics

8. Which of the following sets of numbers has a median of 42?

   F. $\{60, 42, 37, 22, 19\}$
   G. $\{16, 28, 42, 48\}$
   H. $\{42, 64, 20\}$
   J. $\{12, 42, 40, 50\}$
   K. $\{12, 42, 43, 67\}$

9. The student council surveyed the student body on favorite lunch items. The frequency chart below shows the results of the survey.

Favorite Lunch Item	Frequency
corndog	140
hamburger	245
hotdog	210
pizza	235
spaghetti	90
other	65

Which lunch item indicates the mode of the data?

A. other
B. hotdog
C. pizza
D. corndog
E. hamburger

Use the chart for questions 10–11.

Bob	7	4	6	8	7	5	3
Ann	6	5	6	7	9	5	4

10. Find the means. Who has the highest mean?

   F. Ann, 6
   G. Ann, 5.7
   H. Bob, 5.7
   J. Bob, 6
   K. Bob, 9

11. What is the mode for Ann's numbers?

   A. 4
   B. 5
   C. 6
   D. 6 and 5
   E. 9

12. Find the range of the numbers below.

   $14, 52, 20, 68, 37$

   F. 68
   G. 37
   H. 54
   J. 38
   K. 14

13. Find the range of the numbers below.

   $81, 15, 93, 10, 46$

   A. 83
   B. 93
   C. 49
   D. 10
   E. 71

Chapter 22
Data Interpretation

22.1 Tally Charts and Frequency Tables

Large lists can be tallied in a chart. To make a **tally chart**, record a tally mark in a chart for each time a number is listed. To make a **frequency table**, count the times each number occurs in the list, and record the frequency.

Example 1: The age of each student in grades 6–8 in a local middle school are listed below. Make a tally chart and a frequency table for each age.

Student Ages grades 6–8							
10	11	11	12	14	12	11	12
13	13	13	12	14	11	12	11
12	14	12	10	15	11	13	14
12	10	12	11	12	13	12	12
13	12	13	12	11	10	13	11
14	14	11	15	12	13	14	13

TALLY CHART	
Age	Tally
10	IIII
11	HHH HHH
12	HHH HHH HHH
13	HHH HHH
14	HHH II
15	II

FREQUENCY TABLE	
Age	Frequency
10	4
11	10
12	15
13	10
14	7
15	2

Make a chart to record tallies and frequencies for the following problems.

1. The sheriff's office monitors the speed of cars traveling on Turner Road for one week. The following data is the speed of each car that travels Turner Road during the week. Tally the data in 10 mph increments starting with 10–19 mph, and record the frequency in a chart.

Car Speed, mph									
45	52	47	35	48	50	51	43	52	41
40	51	32	24	55	41	32	33	45	
36	39	49	52	34	28	39	47	56	
29	15	63	42	35	42	58	59	35	
39	41	25	34	22	16	40	31	55	
55	10	46	38	50	52	48	36	65	
21	32	36	41	52	49	45	32	20	

Speed	Tally	Frequency
10-19		
20-29		
30-39		
40-49		
50-59		
60-69		

Copyright © American Book Company

259

Chapter 22 Data Interpretation

2. The following data gives final math averages for Ms. Kirby's class. In her class, an average of 90–100 is an A, 80–89 is B, 70–79 is a C, 60–69 is a D, and an average below 60 is an F. Tally and record the frequency of A's, B's, C's D's, and F's.

Final Math Averages									
85	92	87	62	75	84	96	52	31	79
45	77	98	75	71	79	85	82	86	76
87	74	76	68	93	77	65	84	89	
79	65	77	82	86	84	92	60	65	
99	75	88	74	79	80	63	84	69	
87	90	75	81	73	69	73	75	75	

Grade	Tally	Frequency
A		
B		
C		
D		
F		

22.2 Histograms

A **histogram** is a bar graph of the data in a frequency table.

Example 2: Draw a histogram for the customer sales data presented in the frequency table.

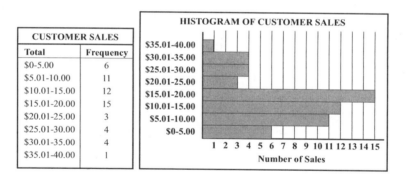

Use the frequency charts that you filled in the previous section to draw histograms for the same data.

1.
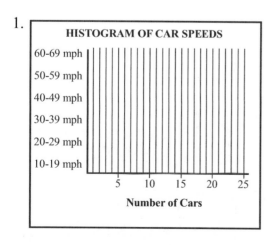

2.

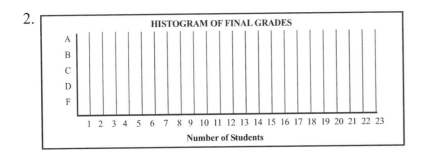

22.3 Reading Tables

A **table** is a concise way to organize large quantities of information using rows and columns.

Read the table carefully, and then answer the questions that follow.

Some employers use a tax table like the one below to calculate how much Federal Income Tax should be withheld from a single person paid weekly. The number of withholding allowances claimed is also commonly referred to as the number of deductions claimed.

Federal Income Tax Withholding Table SINGLE Persons – WEEKLY Payroll Period						
If the wages are –		And the number of withholding allowances claimed is –			
At least	But less than	0	1	2	3
		The amount of income tax to be withheld is –			
$250	260	31	23	16	9
$260	270	32	25	17	10
$270	280	34	26	19	12
$280	290	35	28	20	13
$290	300	37	29	22	15

1. David is single, claims 2 withholding allowances, and earned $275 last week. How much Federal Income Tax was withheld?

2. Cecily is single, claims 0 deductions, and earned $297 last week. How much Federal Income Tax was withheld?

3. Sherri is single, claims 3 deductions, and earned $268 last week. How much Federal Income Tax was withheld from her check?

4. Mitch is single and claims 1 allowance. Last week, he earned $291. How much was withheld from his check for Federal Income Tax?

5. Ginger is single, earns $275 this week, and claims 0 deductions. How much Federal Income Tax is withheld from her check?

6. Bill is single and earns $263 per week. He claims 1 withholding allowance. How much Federal Income Tax is withheld each week?

Chapter 22   Data Interpretation

22.4   Bar Graphs

Bar graphs can be either vertical or horizontal. There may be just one bar or more than one bar for each interval. Sometimes each bar is divided into two or more parts. In this section, you will work with a variety of bar graphs. Be sure to read all titles, keys, and labels to completely understand all the data that is presented. **Answer the questions about each graph.**

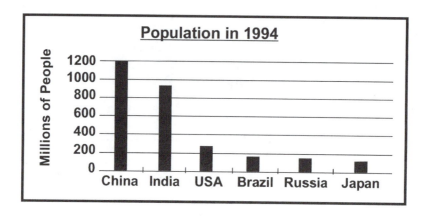

1. Which country has over 1 billion people?

2. How many countries have fewer than 200,000,000 people?

3. How many more people does India have than Japan?

4. If you added together the populations of the USA, Brazil, Russia, and Japan, would the sum be closer to the population of India or China?

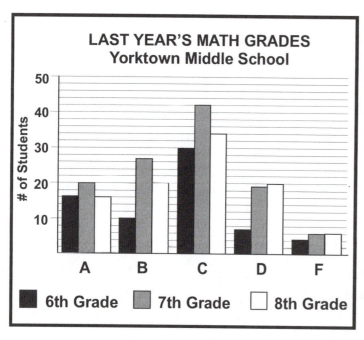

5. How many of last year's 6th graders made C's in math?

6. How many more math students made B's in the 7th grade than in the 8th grade?

7. Which letter grade occurs the most number of times in the data?

8. How many 8th graders took math last year?

9. How many students made A's in math last year?

22.5 Line Graphs

Line graphs often show how data changes over time. Time is always shown on the bottom of the graph. When choosing a graph to show how something changes over time, a line graph is the best choice.

Study the line graphs below, and then answer the questions that follow.

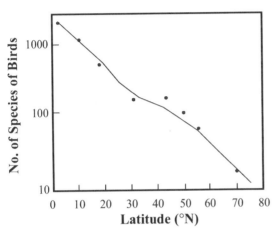
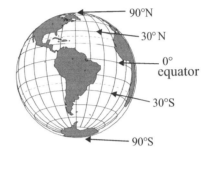

After reading the graph above, label each of the following statements as true or false.

1. There are more species of birds at the North Pole than at the equator.
2. There are more species of birds in Mexico than in Canada.
3. As the latitude increases, the number of species of birds decreases.
4. At 30°N there are over 100 species of birds.
5. The warmer the climate, the fewer kinds of birds there are.

These true or false statements, 6–10, refer to the graph on the left.

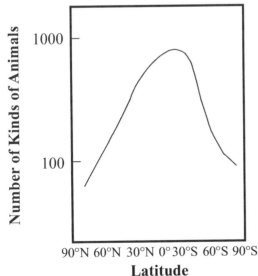

6. The farther north and south you go from the equator, the greater the variety of animals there are.

7. The closer you get to the equator, the greater the variety of animals there are.

8. There are fewer kinds of animals at 30°S than at 60°S latitude.

9. The number of kinds of animals increases as the latitude increases.

10. The number of kinds of animals increases at the poles.

Chapter 22   Data Interpretation

22.6   Multiple Line Graphs

Multiple line graphs are a way to represent a large quantity of data in a small space. It would often take several paragraphs to explain in words the same information that one graph could show.

On the graph below, there are three lines. You will need to read the **key** to understand the meaning of each.

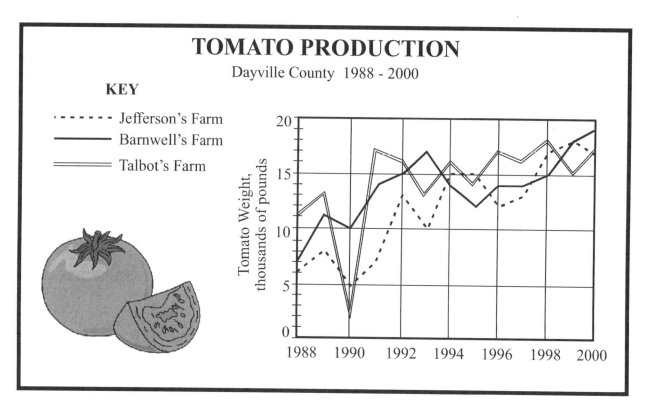

Study the graph, and then answer the questions below.

1. In what year did Barnwell's Farm produce 8,000 pounds of tomatoes more than Talbot's farm?

2. In which year did Dayville County produce the most pounds of tomatoes?

3. In 1993, how many more pounds of tomatoes did Barnwell's Farm produce than Talbot's farm?

4. How many pounds of tomatoes did Dayville County's three farms produce in 1992?

5. In which year did Dayville County produce the fewest pounds of tomatoes?

6. Which farm had the most dramatic increase in production from one year to the next?

7. How many more pounds of tomatoes did Jefferson's Farm produce in 1992 than in 1988?

8. Which farm produced the most pounds of tomatoes in 1995?

22.7 Circle Graphs

Circle graphs represent data expressed in percentages of a total. The parts in a circle graph should always add up to 100%. Circle graphs are sometimes called **pie graphs** or **pie charts**. A few examples of circle graphs are: percentage of kids who prefer a peanut butter and jelly sandwich; or the least favorite color between red, yellow, and purple.

To calculate the value of a percent in a circle graph, multiply the percent by the total. Use the circle graphs below to answer questions. The first question is worked for you as an example.

1. How much does Tina spend each month on music CDs?

   $80 × 0.20 = $16.00

   __$16.00__

2. How much does Tina spend each month on make-up?

   __20.00__

3. How much does Tina spend each month on clothes?

   __40.00__

4. How much does Tina spend each month on snacks?

   __4.00__

Fill in the following chart.

Favorite Activity	Number of Students
5. watching TV	30
6. talking on the phone	15
7. playing video games	28
8. surfing the internet	5
9. playing sports	15
10. reading	7

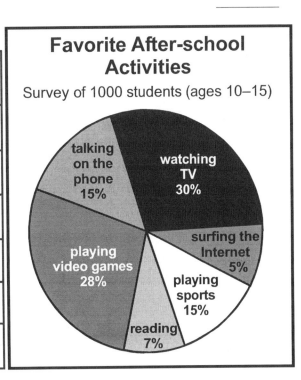

Copyright © American Book Company

Chapter 22   Data Interpretation

$\frac{1}{2} = .5$ or 50%

3/4 = .75 or 75% $\frac{1}{4} = .25$ or 25%

22.8   Pictographs

Pictographs represent data using symbols. The **key** or **legend** tells what the symbol stands for. Before answering questions about the graph, be sure to read the title, key, and horizontal and vertical labels.

Number of Military Officers by Branch of Service

U.S. Bureau of the Census, 1994

NAVY	🦅🦅🦅🦅🦅🦅🦅 = 5,000
MARINES	🌐🌐 3/4 = 75% = 7,500
AIR FORCE	⚪⚪⚪⚪⚪⚪⚪⚪
ARMY	⚪⚪⚪⚪⚪⚪⚪⚪
COAST GUARD	⚓
Key: Each military symbol = 10,000 officers

Answer each question below.

1. How many military officers are in the Marines?

2. Which branch of the service has the fewest number of officers?

3. Are there more Air Force officers or Army officers?

4. How many military officers are there in all?

5. How many officers do the Navy and Marines have altogether?

6. How many more Navy officers are there than Marine officers?

7. How many Army and Coast Guard officers are there altogether?

266   Copyright © American Book Company

22.9 Reading Venn Diagrams

Venn diagrams are a visual way to see two or more variables. They show whether or not the variables intersect. A Venn diagram also shows the union of the events.

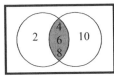

$\{2, 4, 6, 8\} \cap \{4, 6, 8, 10\} = \{4, 6, 8\}$
The shaded area is the intersection of the two sets.

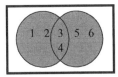

$\{1, 2, 3, 4\} \cup \{3, 4, 5, 6\} = \{1, 2, 3, 4, 5, 6\}$
The shaded area is the union of the two sets.

Example 3: Below is a Venn diagram of how many students play football and baseball. Find the intersection and the union of the two events below.

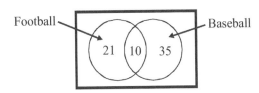

Step 1: First, you must figure out how many students play each sport by interpreting the diagram. Ten students play both sports, since they are counted on the football and baseball side. **The intersection is 10 students.**

Step 2: To find the union, you must add all the players together.
The union is $21 + 35 + 10 = 66$ **students**.

Look at the Venn diagram to find the following.

1. Dogs ∩ Cats
2. Dogs ∪ Birds
3. Cats ∩ Birds
4. Dogs ∩ Cats ∩ Birds
5. Dogs ∪ Cats
6. Dogs ∪ Cats ∪ Birds
7. Dogs ∩ Birds
8. Cats ∪ Birds
9. Dogs ∪ (Cats ∪ Birds)
10. (Dogs ∩ Cats) ∩ Birds

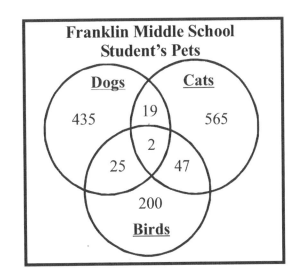

Chapter 22 Review

Use the data given to answer the questions that follow.

The 6th grade did a survey on the number of pets each student had at home. The following give the data produced by the survey.

NUMBER OF PETS PER STUDENT
0 2 6 2 1 0 4 2 3 3 0 2 3 5 1 4 2 0 5 2 3 3 4 3 6 2
5 1 2 3 5 6 3 2 2 5 2 3 4 3 0 1 4 1 2 4 5 7 6 1 4 7

1. Fill in the frequency table.

Number of Pets	Frequency

2. Fill in the histogram.

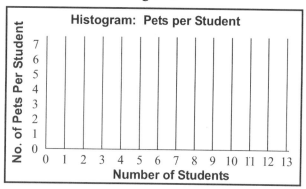

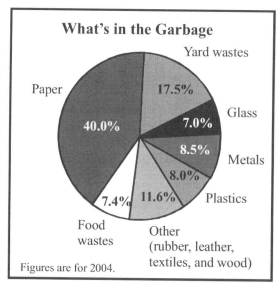

KNIGHTS BASKETBALL Points Scored				
Player	game 1	game 2	game 3	game 4
Joey	5	2	4	8
Jason	10	8	10	12
Brandon	2	6	5	6
Ned	1	3	6	2
Austin	0	4	7	8
David	7	2	9	4
Zac	8	6	7	4

3. How many points did the Knights basketball team score in game 1?

4. How many more points does David score in game 3 than in game 1?

5. How many points does Jason score in the first 4 games?

6. In 2004, the United States produced 160 million metric tons of garbage. According to the pie chart, how much glass was in the garbage?

7. Out of the 160 million metric tons of garbage, how much was glass, plastic, and metal?

8. If in 2007, the garbage reaches 200 million metric tons, and the percentage of wastes remains the same as in 2004, how much food in metric tons will be in the 2007 garbage?

Chapter 22 Review

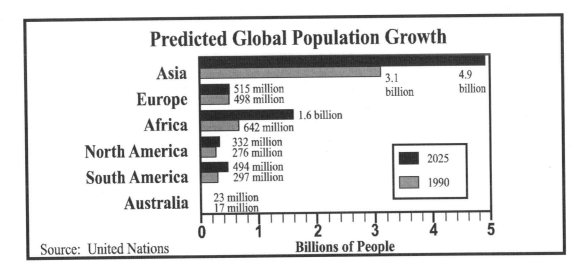

9. By how many is Asia's population predicted to increase between 1990 and 2025?

10. In 1990, how much larger was Africa's population than Europe's?

11. Where is the population expected to more than double between 1990 and 2025?

Use the following Venn diagram to answer questions 17 and 18.

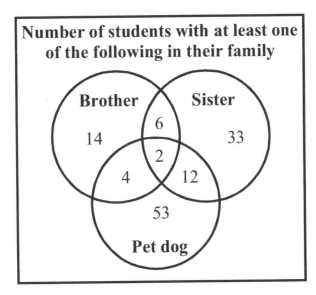

12. In the Venn diagram above, how many members are in the following set?
{sister ∪ pet dog}

13. In the Venn diagram above, how many members are in the following set?
{brother} ∩ {sister}

Copyright © American Book Company   269

Chapter 22 Data Interpretation

Chapter 22 Test

Read the table below, and answer questions 1 and 2.

Name	Total CDs owned
Maggie	97
Erica	164
John	81
Philip	151
Tanya	122

1. Which person has about twice as many CDs as John?

   A. Philip
   B. Tanya
   C. Erica
   D. Maggie
   E. John

2. Which person owns about 20% less than Philip?

   F. Maggie
   G. Phillip
   H. John
   J. Erica
   K. Tanya

Use the line graph below, and answer questions 3 and 4.

3. Which two months marked the greatest increase in sales?

   A. May – June
   B. June – July
   C. March – April
   D. April – May
   E. March – May

4. If the owner expects sales in August to be 10% higher than July, how much should sales be in August?

   F. $902
   G. $750
   H. $792
   J. $800
   K. $782

Chapter 22 Test

Use the circle graph below, and answer questions 5 and 6.

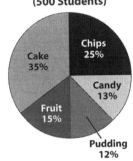

5. If 500 students were surveyed, how many students prefer cake as a favorite snack?

   A. 450 students
   B. 130 students
   C. 150 students
   D. 195 students
   E. 175 students

6. How many students prefer candy and pudding?

   F. 65 students
   G. 60 students
   H. 125 students
   J. 100 students
   K. 115 students

Use the Venn diagram below to answer questions 7–8.

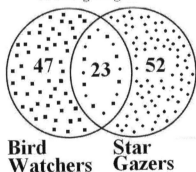

7. What is the total number of bird watchers ∪ star gazers?

   A. 47
   B. 23
   C. 52
   D. 122
   E. 5

8. What is the total number of bird watchers ∩ star gazers?

   F. 47
   G. 23
   H. 52
   J. 122
   K. 5

Copyright © American Book Company

Chapter 22 Data Interpretation

9. The double line graph below shows the number of pages Joshua and Jeremy read.

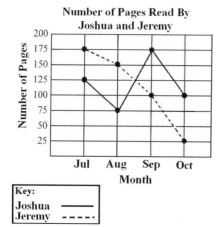

Which statement is true about the data in this line graph?

A. Joshua read less every month than Jeremy.
B. Joshua read less than Jeremy during two months.
C. Jeremy read less every month than Jeremy.
D. Jeremy read more every month than Jeremy.
E. Jeremy and Joshua read the same number of pages over the four months.

10. The double bar graph below shows the favorite kinds of music in Mr. Klunoa's music class.

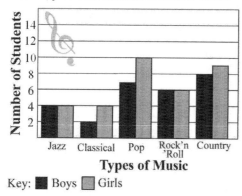

Which of the pairs of statements are true about the data in the graph?

F. Most of the girls prefer classical music. Most of the boys prefer country.
G. The same number of boys and girls said they like jazz. Less girls like country than boys who like country.
H. The same number of boys and girls like rock'n'roll. More girls like pop music than boys like pop music.
J. More girls like country than boys like country. More boys like classical than girls like classical.
K. More boys like pop music than girls like pop music. An equal number of boys and girls like jazz music.

Chapter 23
Probability

23.1 Probability

Probability is the chance something will happen. Probability is most often expressed as a fraction, a decimal, a percent, or can also be written out in words.

Example 1: Billy has 3 red marbles, 5 white marbles, and 4 blue marbles on the floor. His cat comes along and bats one marble under the chair. What is the **probability** it is a red marble?

Step 1: The number of red marbles will be the top number of the fraction. ⟶ $\dfrac{3}{12}$

Step 2: The total number of marbles is the bottom number of the fraction.

The answer may be expressed in lowest terms. $\dfrac{3}{12} = \dfrac{1}{4}$.

Expressed as a decimal, $\tfrac{1}{4} = 0.25$, as a percent, $\tfrac{1}{4} = 25\%$, and written out in words $\tfrac{1}{4}$ is "one out of four."

Example 2: Determine the probability that the pointer will stop on a shaded wedge or the number 1.

Step 1: Count the number of possible wedges that the spinner can stop on to satisfy the above problem. There are 5 wedges that satisfy it (4 shaded wedges and one number 1). The top number of the fraction is 5.

Step 2: Count the total number of wedges, 7. The bottom number of the fraction is 7. The answer is $\tfrac{5}{7}$ or "five out of seven."

Example 3: Refer to the spinner in the example above. If the pointer stops on the number 7, what is the probability that it will **not** stop on 7 the next time?

Step 1: If P is the probability of an event occurring, $1 - P$ is the probability of an event **not** occurring (it is the complement). In this example, the probability of the spinner landing on 7 is $\tfrac{1}{7}$.

Step 2: The probability that the spinner will not stop on 7 is $1 - \tfrac{1}{7}$ which equals $\tfrac{6}{7}$. The answer is $\tfrac{6}{7}$ or "six out of seven."

Copyright © American Book Company

Chapter 23 Probability

Find the probability of the following problems. Express the answer as a percent.

1. A computer chooses a random number between 1 and 50. What is the probability that you will guess the same number that the computer chose in 1 try?

2. There are 24 candy-coated chocolate pieces in a bag. Eight have defects in the coating that can be seen only with close inspection. What is the probability of pulling out a defective piece without looking?

3. Seven sisters have to choose which day each will wash the dishes. They put equal-sized pieces of paper in a hat, each labeled with a day of the week. What is the probability that the first sister who draws will choose a weekend day?

4. For his garden, Clay has a mixture of 12 white corn seeds, 24 yellow corn seeds, and 16 bicolor corn seeds. If he reaches for a seed without looking, what is the probability that Clay will plant a bicolor corn seed first?

5. Carla just got a new department store credit card in the mail. What is the probability that the last digit is an odd number?

6. Alex has a paper bag of cookies that holds 8 chocolate chip, 4 peanut butter, 6 butterscotch chip, and 12 ginger. Without looking, his friend John reaches in the bag for a cookie. What is the probability that the cookie is peanut butter?

7. An umpire at a little league baseball game has 14 balls in his pockets. Five of the balls are brand A, 6 are brand B, and 3 are brand C. What is the probability that the next ball he throws to the pitcher is a brand C ball?

8. What is the probability that the spinner's arrow will land on an even number?

9. The spinner in the problem above stopped on a shaded wedge on the first spin and stopped on the number 2 on the second spin. What is the probability that it will not stop on a shaded wedge or on the 2 on the third spin?

10. A company is offering 1 grand prize, 3 second place prizes, and 25 third place prizes based on a random drawing of contest entries. If your entry is one of the 500 total entries, what is the probability you will win a third place prize?

23.2   More Probability

23.2   More Probability

Example 4:   You have a cube with one number, 1, 2, 3, 4, 5, or 6 painted on each face of the cube. What is the probability that if you throw the cube 3 times, you will get the number 2 each time?

If you roll the cube once, you have a 1 in 6 chance of getting the number 2. If you roll the cube a second time, you again have a 1 in 6 chance of getting the number 2. If you roll the cube a third time, you again have a 1 in 6 chance of getting the number 2. The probability of rolling the number 2 three times in a row is:

$$\frac{1}{6} \times \frac{1}{6} \times \frac{1}{6} = \frac{1}{216}$$

Find the probability that each of the following events will occur.

There are 10 balls in a box, each with a different digit on it: 0, 1, 2, 3, 4, 5, 6, 7, 8, & 9. A ball is chosen at random and then put back in the box.

1. What is the probability that if you pick out a ball 3 times, you will get number 7 each time?

2. What is the probability you will pick a ball with 5, then 9, and then 3?

3. What is the probability that if you pick out a ball 4 times, you will always get an odd number?

4. A couple has 4 children ages 9, 6, 4, and 1. What is the probability that they are all girls?

There are 26 letters in the alphabet, allowing a different letter to be on each of 26 cards. The cards are shuffled. After each card is chosen at random, it is put back in the stack of cards, and the cards are shuffled again.

5. What is the probability that when you pick 3 cards, you would draw first a "y", then an "e", and then an "s"?

6. What is the probability that you would draw 4 cards and get the letter "z" each time?

7. What is the probability that you draw twice and get a letter in the word "random" both times?

8. If you flip a coin 3 times, what is the probability you will get heads every time?

9. Marie is clueless about 4 of her multiple-choice answers. The possible answers are A, B, C, D, E, or F. What is the probability that she will guess all four answers correctly?

Copyright © American Book Company

275

Chapter 23   Probability

23.3   Tree Diagrams

Drawing a tree diagram is another method of determining the probability of events occurring.

Example 5:   If you toss two six-sided numbered cubes that have 1, 2, 3, 4, 5, or 6 on each side, what is the probability you will get two cubes that add up to 9? One way to determine the probability is to make a tree diagram.

Cube 1	Cube 2	Cube 1 plus Cube 2
1	1	2
	2	3
	3	4
	4	5
	5	6
	6	7
2	1	3
	2	4
	3	5
	4	6
	5	7
	6	8
3	1	4
	2	5
	3	6
	4	7
	5	8
	6	⑨
4	1	5
	2	6
	3	7
	4	8
	5	⑨
	6	10
5	1	6
	2	7
	3	8
	4	⑨
	5	10
	6	11
6	1	7
	2	8
	3	⑨
	4	10
	5	11
	6	12

Alternative method

Write down all of the numbers on both cubes which would add up to 9.

Cube 1	Cube 2
4	5
5	4
6	3
3	6

Numerator = 4 combinations

For denominator: Multiply the number of sides on one cube times the number of sides on the other cube.

$6 \times 6 = 36$

Numerator:
Denominator: $\dfrac{4}{36} = \dfrac{1}{9}$

There are 36 possible ways the cubes could land. Out of those 36 ways, the two cubes add up to 9 only 4 times. The probability you will get two cubes that add up to 9 is $\dfrac{4}{36}$ or $\dfrac{1}{9}$.

276   Copyright © American Book Company

23.3   Tree Diagrams

Read each of the problems below. Then answer the questions.

1. Jake has a spinner. The spinner is divided into eight equal regions numbered 1–8. In two spins, what is the probability that the numbers added together will equal 12?

2. Charlie and Libby each spin one spinner one time. The spinner is divided into 5 equal regions numbered 1–5. What is the probability that these two spins added together would equal 7?

3. Gail spins a spinner twice. The spinner is divided into 9 equal regions numbered 1–9. In two spins, what is the probability that the difference between the two numbers will equal 4?

4. Diedra throws two 10-sided numbered polyhedrons. What is the probability that the difference between the two numbers will equal 7?

5. Cameron throws two six-sided numbered cubes. What is the probability that the difference between the two numbers will equal 3?

6. Tesla spins one spinner twice. The spinner is divided into 11 equal regions numbered 1–11. What is the probability that the two numbers added together will equal 11?

7. Samantha decides to roll two five-sided numbered prisms. What is the probability that the two numbers added together will equal 4?

8. Mary Ellen spins a spinner twice. The spinner is divided into 7 equal regions numbered 1–7. What is the probability that the product of the two numbers equals 10?

9. Conner decides to roll two six-sided numbered cubes. What is the probability that the product of the two numbers equals 4?

10. Tabitha spins one spinner twice. The spinner is divided into 9 equal regions numbered 1–9. What is the probability that the sum of the two numbers equals 10?

11. Darnell decides to roll two 15-sided numbered polyhedrons. What is the probability that the difference between the two numbers is 13?

12. Inez spins one spinner twice. The spinner is divided into 12 equal regions numbered 1–12. What is the probability that the sum of two numbers equals 10?

13. Gina spins one spinner twice. The spinner is divided into 8 equal regions numbered 1–8. What is the probability that the two numbers added together equals 9?

14. Celia rolls two six-sided numbered cubes. What is the probability that the difference between the two numbers is 2?

15. Brett spins one spinner twice. The spinner is divided into 4 equal regions numbered 1–4. What is the probability that the difference between the two numbers will be 3?

Copyright © American Book Company

277

Chapter 23   Probability

23.4   Independent and Dependent Events

In mathematics, the outcome of an event may or may not influence the outcome of a second event. If the outcome of one event does not influence the outcome of the second event, these events are **independent.** However, if one event has an influence on the second event, the events are **dependent**. When someone needs to determine the probability of two events occurring, he or she will need to use an equation. These equations will change depending on whether the events are independent or dependent in relation to each other. When finding the probability of two **independent** events, multiply the probability of each favorable outcome together. Independent events use the **multiplication rule**. The multiplication rule for independent events is $P(A \text{ and } B) = P(A) P(B)$.

Example 6:   One bag of marbles contains 1 white, 1 yellow, 2 blue, and 3 orange marbles. A second bag of marbles contains 2 white, 3 yellow, 1 blue, and 2 orange marbles. What is the probability of drawing a blue marble from each bag?

Solution: Probability of favorable outcomes

Bag 1 $P(A)$: $\dfrac{2}{7}$

Bag 2 $P(B)$: $\dfrac{1}{8}$

Probability of a blue marble from each bag $= P(A \text{ and } B)$

$$P(A \text{ and } B) = P(A) P(B) = \frac{2}{7} \times \frac{1}{8} = \frac{2}{56} = \frac{1}{28}$$

In order to find the probability of two **dependent** events, you will need to use a different set of rules. For the first event, you must divide the number of favorable outcomes by the number of possible outcomes. For the second event, you must subtract one from the number of favorable outcomes **only if** the favorable outcome is the **same**. However, you must subtract one from the number of total possible outcomes. Finally, you must multiply the probability for event one by the probability for event two.

Example 7:   One bag of marbles contains 3 red, 4 green, 7 black, and 2 yellow marbles. What is the probability of drawing a green marble, removing it from the bag, and then drawing another green marble without looking?

	Favorable Outcomes	Total Possible Outcomes
Draw 1	4	16
Draw 2	3	15
Draw 1 $\times$ Draw 2	12	240

Answer: $\dfrac{12}{240}$ or $\dfrac{1}{20}$

Example 8:   Using the same bag of marbles, what is the probability of drawing a red marble and then drawing a black marble?

	Favorable Outcomes	Total Possible Outcomes
Draw 1	3	16
Draw 2	7	15
Draw 1 $\times$ Draw 2	21	240

Answer $\dfrac{21}{240}$ or $\dfrac{7}{80}$

278   Copyright © American Book Company

23.4   Independent and Dependent Events

Find the probability of the following problems. Express the answer as a fraction.

1. Prithi has two boxes. Box 1 contains 3 red, 2 silver, 4 gold, and 2 blue combs. She also has a second box containing 1 black and 1 clear brush. What is the probability that Prithi selects a red comb from box 1 and a black brush from box 2?

2. Steve Marduke has two spinners in front of him. The first one is numbered 1–6, and the second is numbered 1–3. If Steve spins each spinner once, what is the probability that the first spinner will show an odd number and the second spinner will show a "1"?

3. Carrie McCallister flips a coin twice and gets heads both times. What is the probability that Carrie will get tails the third time she flips the coin?

4. Artie Drake turns a spinner which is evenly divided into 11 sections numbered 1–11. On the first spin, Artie's pointer lands on "8". What is the probability that the spinner lands on an even number the second time he turns the spinner?

5. Leanne Davis plays a game with a street entertainer. In this game, a ball is placed under one of three coconut halves. The vendor shifts the coconut halves so quickly that Leanne can no longer tell which coconut half contains the ball. She selects one and misses. The entertainer then shifts all three around once more and asks Leanne to pick again. What is the probability that Leanne will select the coconut half containing the ball?

6. What is the probability that Jane Robelot reaches into a bag containing 1 daffodil and 2 gladiola bulbs and pulls out a daffodil bulb, and then reaches into a second bag containing 6 tulip, 3 lily, and 2 gladiola bulbs and pulls out a lily bulb?

7. Terrell casts his line into a pond containing 7 catfish, 8 bream, 3 trout, and 6 northern pike. He immediately catches a bream and keeps it. What are the chances that Terrell will catch a second bream the next time he casts his line?

8. Gloria Quintero enters a contest in which the person who draws his or her initials out of a box containing all 26 letters of the alphabet wins the grand prize. Gloria reaches in, draws a "G", keeps it, then draws another letter. What is the probability that Gloria will next draw a "Q"?

9. Vince Macaluso is pulling two socks out of a washing machine in the dark. The washing machine contains three tan, one white, and two black socks. If Vince reaches in and pulls out the socks one at a time, what is the probability that he will pull out two tan socks on his first two tries?

10. John Salome has a bag containing 2 yellow plums, 2 red plums, and 3 purple plums. What is the probability that he reaches in without looking and pulls out a yellow plum and eats it, then reaches in again without looking and pulls out a red plum to eat?

Copyright © American Book Company

279

Chapter 23 Probability

23.5 Permutations

A **permutation** is an arrangement of items in a specific order. If a problem asks how many ways can you arrange 6 books on a bookshelf, it is asking how many permutations there are for 6 items.

Example 9: Ron has 4 items: a model airplane, a trophy, and autographed football, and a toy sports car. How many ways can he arrange the 4 items on a shelf?

Solution: Ron has 4 choices for the first item on the shelf. He then has 3 choices left for the second item. After choosing the second item, he has 2 choices left for the third item and only one choice for the last item. The diagram below shows the permutations for arranging the 4 items on a shelf if he chose to put the trophy first.

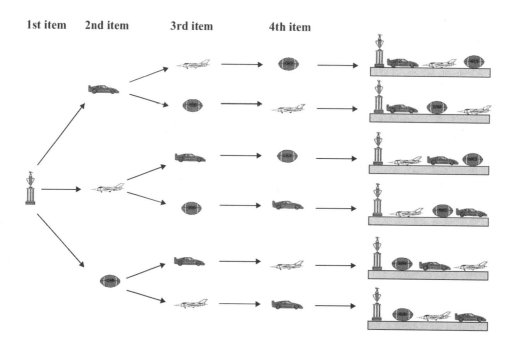

Count the number of permutations if Ron chooses the trophy as the first item. There are 6 permutations. Next, you could construct a pyramid of permutations choosing the model car first. That pyramid would also have 6 permutations. Then, you could construct a pyramid choosing the airplane first. Finally, you could construct a pyramid choosing the football first. You would then have a total of 4 pyramids each having 6 permutations. The total number of permutations is $6 \times 4 = 24$. There are 24 ways to arrange the 4 items on a bookshelf.

You probably don't want to draw pyramids for every permutation problem. What if you want to know the permutations for arranging 30 objects? Fortunately, mathematicians have come up with a formula for calculating permutations.

For the above problem, Ron has 4 items to arrange. Therefore, multiply $4 \times 3 \times 2 \times 1 = 24$. Another way of expressing this calculation is 4!, stated as 4 factorial. $4! = 4 \times 3 \times 2 \times 1$.

23.5   Permutations

Example 10:   How many ways can you line up 6 students?

Solution: The number of permutations for 6 students $= 6! = 6 \times 5 \times 4 \times 3 \times 2 \times 1 = 720$. There are 6 choices for the first position, 5 for the second position, 4 for the third, 3 for the fourth, 2 for the fifth, and 1 for the sixth.

Example 11:   Shelley and her mom, dad, and brother are having cake for her birthday. Since it is Shelley's birthday she gets a piece first. How many ways are there to pass out the pieces of cake?

Solution: Since Shelley gets the first piece, the first spot is fixed. The second, third, and fourth spots are not fixed and anyone left can be in one of the three spots.

Spot	1	2	3	4
Choices of people	1	3	2	1

Now, multiply the choices together, $1 \times 3 \times 2 \times 1 = 6$ ways to pass out cake.

Work the following permutation problems.

1. How many ways can you arrange 6 books on a bookshelf?

2. Myra has 5 novels to arrange on a book shelf. How many ways can she arrange the novels?

3. Four sprinters signed up for the 100 meter dash. How many ways can the four sprinters line up on the starting line?

4. Keri wants an ice cream cone with one scoop of chocolate, one scoop of vanilla, and one scoop of strawberry. How many ways can the scoops be arranged on the cone?

5. How many ways can you arrange the letters A, B, C, D, E, F, and G?

6. At Sam's party, the DJ has three song requests. In how many different orders can he play the 3 songs?

7. Yvette has 6 comic books. How many different ways can she stack the comic books?

8. Sandra's couch can hold four people. How many ways can she and her three friends sit on the couch?

9. How many ways can you arrange the numbers 2, 3, 5, 6?

10. At a busy family restaurant, five tables open up at the same time. How many different ways can the hostess seat the next five families waiting to be seated?

11. How many ways can you arrange the numbers 1, 2, 3, 4, 5, 6, 7, 8, and always have 3 at position 1 and 5 at position 5?

Copyright © American Book Company

281

Chapter 23   Probability

23.6   More Permutations

Example 12:   If there are 6 students, how many ways can you line up any 4 of them?

Solution: Multiply $6 \times 5 \times 4 \times 3 = 360$. There are 6 choices for the first position in line, 5 for the second position, 4 for the third position, and 3 for the last position. There are 360 ways to line up 4 of the 6 students.

Find the number of permutations for each of the problems below.

1. How many ways can you arrange 3 out of 8 books on a shelf?

2. How many 4 digit numbers can be made using the numbers 2, 3, 5, 8, and 9?

3. How many ways can you line up 3 students out of a class of 20?

4. Kim worked in the linen department of a store. Six new colors of towels came in. Her job was to line up the new towels on a long shelf. How many ways can she arrange the 6 colors?

5. Terry's CD player holds 4 CDs. Terry owns 10 CDs. How many different ways can he arrange his CDs in the CD player?

6. Erik has 8 shirts he wears to school. How many ways can he choose a different shirt to wear on Monday, Tuesday, Wednesday, Thursday, and Friday?

7. Deb has a box of 8 markers. The art teacher told her to choose two markers and line them up on her desk. How many ways can she line up 2 markers from the 8?

8. Jeff went into an ice cream store serving 16 flavors of ice cream. He wanted a cone with two different flavors. How many ways could he order 2 scoops of ice cream, one on top of the other?

9. In how many ways can you arrange any 2 letters from the 26 letters in the alphabet?

282   Copyright © American Book Company

23.7 Combinations

In a **permutation**, objects are arranged in a particular order. In a **combination**, the order does not matter. In a **permutation**, if someone picked two letters of the alphabet, **k, m** and **m, k** would be considered 2 different permutations. In a **combination**, **k, m** and **m, k** would be the same combination. A different order does not make a new combination.

Example 13: How many combinations of 3 letters from the set {a, b, c, d, e} are there?

Step 1: Find the **permutation** of 3 out of 5 objects.

Step 2: Divide by the permutation of the **number of objects** to be chosen from the total (3). This step eliminates the duplicates in finding the permutations.

Step 3: Cancel common factors and simplify.

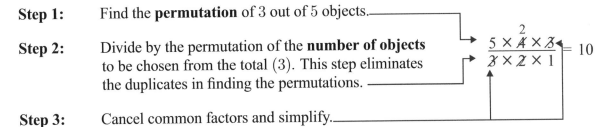

There can be 10 combinations of three letters from the set {a, b, c, d, e}.

Find the number of combinations for each problem below.

1. How many combinations of 3 numbers can be made from the set of numbers {2, 4, 6, 7, 8, 9}?

2. Jackson Middle School wants to choose 3 students at random from the 8th grade to take an opinion poll. There are 100 eighth graders in the school. How many different groups of 3 students could be chosen? (Use a calculator for this one.)

3. How many combinations of 2 students can be made from a class of 10?

4. Fashion Ware catalog has a sweater that comes in 6 colors. How many combinations of 2 different colors does a shopper have to choose from?

5. Angelo's Pizza offers 12 different pizza toppings. How many different combinations can be made of pizzas with three toppings?

6. How many different combinations of 4 flavors of jelly beans can you make from a store that sells 20 different flavors of jelly beans?

7. The track team is running the relay race in a competition this Saturday. There are 8 members of the track team. The relay race requires 2 runners. How many combinations of 2 runners can be formed from the track team?

8. Keisha got to pick 2 prizes from a grab bag containing 10 prizes. How many combinations of 2 prizes are possible?

Chapter 23   Probability

23.8   More Combinations

Another kind of combination involves selection from several categories.

Example 14:   At Joe's Deli, you can choose from 4 kinds of bread, 5 meats, and 3 cheeses when you order a sandwich. How many different sandwiches can be made with Joe's choices for breads, meats, and cheeses if you choose 1 kind of bread, 1 meat, and 1 cheese for each sandwich?

JOE'S SANDWICHES

Breads	**Meats**	**Cheeses**
White	Roast beef	Swiss
Pumpernickel	Corned beef	American
Light rye	Pastrami	Mozzarella
Whole wheat	Roast chicken	
	Roast turkey	

Solution: Multiply the number of choices in each category. There are 4 breads, 5 meats, and 3 cheeses, so $4 \times 5 \times 3 = 60$. There are 60 combinations of sandwiches.

Find the number of combinations that can be made in each of the problems below.

1. Drew has 10 pairs of socks, 5 shirts, and 3 pairs of tennis shoes. How many different outfit combinations can be made with Drew's clothes?

2. Tyler has 5 baseball caps, 3 jackets, 5 pairs of jeans, and 2 pairs of sunglasses. How many combinations of the 4 items can he make?

3. Amber has 4 kinds of lipstick, 8 eye shadows, 3 kinds of lip liner, and 1 mascara. How many combinations can she use to make up her face?

4. Clarence's dad is ordering a new car. He has a choice of 4 exterior colors, 2 interior colors, 2 kinds of seats, and 3 sound systems. How many combinations does he have to pick from?

5. A fast food restaurant has 6 kinds of sandwiches, 2 kinds of French fries, and 8 kinds of soft drinks. How many combinations of meals could you order if you ordered a sandwich, fries, and a drink?

6. In summer camp, Jake can choose from 5 outdoor activities, 2 indoor activities, and 3 water sports. He has to choose one of each. How many combinations of activities can he choose?

7. Ella won a contest at school and gets to choose one pencil and one pen from the school store and a pizza from the lunch room. There are 5 colors of pencils, 3 colors of pens, and 2 kinds of pizza. How many combinations of prize packages can she choose?

284   Copyright © American Book Company

Chapter 23 Review

1. There are 50 students in the school orchestra in the following sections:

string section	woodwind	percussion	brass
25	15	5	5

   One student will be chosen at random to present the orchestra director with an award. What is the probability the student will be from the woodwind section?

2. Fluffy's cat treat box contains 6 chicken-flavored treats, 5 beef-flavored treats, and 7 fish-flavored treats. If Fluffy's owner reaches in the box without looking, and chooses one treat, what is the probability that Fluffy will get a chicken-flavored treat?

3. The spinner in figure A stopped on the number 5 on the first spin. What is the probability that it will not stop on 5 on the second spin?

Fig. A   Fig. B

4. Sherri turns the spinner in figure B above 3 times. What is the probability that the pointer always lands on a shaded number?

5. Three cakes are sliced into 20 pieces each. Each cake contains 1 gold ring. What is the probability that one person who eats one piece of cake from each of the 3 cakes will find 3 gold rings?

6. Brianna tosses a coin 4 times. What is the probability she gets all tails?

7. Tempest has a bag with 4 red marbles, 3 blue marbles, and 2 yellow marbles. Does adding 4 purple marbles increase or decrease her chances that the first marble she draws at random will be red?

8. Simone has lived in Silver Spring for 250 days, and in that time her power has been out 3 days.
   (A) If the power outages happen at random, what is the probability that the power will be out tomorrow?
   (B) If the probability remains the same, how many days will she be without power for the next 10 years (3,652 days)? Round to the nearest whole number.

Read the following, and answer questions 9–13.

There are 9 slips of paper in a hat, each with a number from 1 to 9. The numbers correspond to a group of students who must answer a question when the number for their group is drawn. Each time a number is drawn, the number is put back in the hat.

9. What is the probability that the number 6 will be drawn twice in a row?

10. What is the probability that the first 5 numbers drawn will be odd numbers?

11. What is the probability that the second, third, and fourth numbers drawn will be even numbers?

12. What is the probability that the first five times a number is drawn it will be the number 5?

13. What is the probability that the first five numbers drawn will be 1, 2, 3, 4, 5 in that order?

Chapter 23   Probability

Solve the following word problems. For questions 14–16, write whether the problem is "dependent" or "independent".

14. Felix Perez reaches into a 10-piece puzzle and pulls out one piece at random. This piece has two places where it could connect to other pieces. What is the probability that he will select another piece which fits the first one if he selects the next piece at random?

15. Barbara Stein is desperate for a piece of chocolate candy. She reaches into a bag which contains 8 peppermint, 5 butterscotch, 7 toffee, 3 mint, and 6 chocolate pieces and pulls out a toffee piece. Disappointed, she throws it back into the bag and then reaches back in and pulls out one piece of candy. What is the probability that Barbara pulls out a chocolate piece on the second try?

16. Christen Solis goes to a pet shop and immediately decides to purchase a guppy she saw swimming in an aquarium. She reaches into the tank containing 5 goldfish, 6 guppies, 4 miniature catfish, and 3 minnows and accidently pulls up a goldfish. Breathing a sigh, Christen places the goldfish back in the water. The fish are swimming so fast, it is impossible to tell what fish Christen would catch. What is the probability that Christen will catch a guppy on her second try?

17. Julia has 7 porcelain statues. How many ways can she arrange 3 of the statues on a display shelf?

18. Ms. White has 8 students. Every day she randomly draws names of 2 students out of a bag to turn in their homework for a test grade. How many combinations of 2 students can she draw?

19. In the lunch line, students can choose 1 out of 2 meats, 1 out of 3 vegetables, 1 out of 3 desserts, and 1 out of 4 drinks. How many lunch combinations are there?

286   Copyright © American Book Company

Chapter 23 Test

1. There are 10 boys and 12 girls in a class. If one student is selected at random from the class, what is the probability it is a girl?

   A. $\frac{1}{2}$

   B. $\frac{1}{22}$

   C. $\frac{6}{11}$

   D. $\frac{6}{5}$

   E. $\frac{1}{12}$

2. David just got a new credit card in the mail. What is the probability the second digit of the credit card number is a 3?

   F. $\frac{1}{3}$

   G. $\frac{1}{9}$

   H. $\frac{2}{3}$

   J. $\frac{1}{5}$

   K. $\frac{1}{10}$

3. Brenda has 18 fish in an aquarium. The fish are the following colors: 5 orange, 7 blue, 2 black, and 4 green. Brenda also has a trouble-making cat that has grabbed a fish. What is the probability the cat grabbed a green fish if all the fish are equally capable of avoiding the cat?

   A. $\frac{2}{9}$

   B. $\frac{1}{18}$

   C. $\frac{2}{7}$

   D. $\frac{1}{4}$

   E. $\frac{7}{9}$

4. In problem number 3, what is the probability the cat will **not** grab an orange fish?

   F. $\frac{1}{3}$

   G. $\frac{13}{18}$

   H. $\frac{3}{4}$

   J. $\frac{13}{5}$

   K. $\frac{1}{18}$

5. You have a cube with each face numbered 1, 2, 3, 4, 5, or 6. What is the probability if you roll the cube 4 times, you will get the number 5 each time?

   A. $\frac{4}{1296}$

   B. $\frac{4}{5}$

   C. $\frac{1}{256}$

   D. $\frac{1}{1296}$

   E. $\frac{2}{3}$

6. In a game using two numbered cubes, what is the probability of **not** rolling the same number on both cubes three times in a row?

   F. $\frac{125}{256}$

   G. $\frac{125}{216}$

   H. $\frac{27}{64}$

   J. $\frac{16}{27}$

   K. $\frac{1}{36}$

7. Carrie bought a large basket of 60 apples. When she got them home, she found 4 of the apples were rotten. If she goes back and buys 200 more apples, about how many rotten apples should she expect?

   A. 8

   B. 15

   C. 20

   D. 24

   E. 13

Copyright © American Book Company

Chapter 23   Probability

8. Katie spun a spinner 15 times and recorded her results in the table below. The spinner was divided into 6 sections numbered 1–6.

$$
\begin{array}{ccccc}
1 & 3 & 6 & 5 & 1 \\
6 & 2 & 4 & 3 & 4 \\
2 & 5 & 1 & 4 & 5 \\
\end{array}
$$

Based on the results, how many times would 4 be expected to appear in 45 spins?

F. 9
G. 12
H. 15
J. 21
K. 24

9. Sarah has a bag of 100 marbles. 40 are blue, 40 are green, and 20 are purple. What is the probability of Sarah picking a purple marble, without looking, on the first try?

A. $\frac{1}{5}$

B. $\frac{1}{4}$

C. $\frac{1}{100}$

D. $\frac{8}{25}$

E. $\frac{1}{10}$

10. Kyle has 4 different kinds of candy in a bag. He has 5 chocolate rolls, 6 lollipops, 4 chocolate bars, and 7 peanut butter candies. What kind of candy must Kyle have picked out of the bag if the probability of picking that kind is $\frac{2}{11}$?

F. chocolate roll
G. lollipop
H. all 4 kinds
J. pcanut butter
K. chocolate bar

11. There are 7 sections of equal size on a spinner. One is labeled purple. Two are labeled green. Three are labeled orange. One section is unlabeled. What color must the unlabeled section be if the probability of spinning that color is $\frac{4}{7}$?

A. purple
B. green
C. orange
D. brown
E. unlabeled

12. The Toy Tractor Company has trouble with the wheels coming off some of the toys. A random sample of 60 toys find that the wheels fell off 12 of the toys. It has produced 640 of these toys. About how many of them could they expect to find defective wheels?

F. 12
G. 60
H. 628
J. 128
K. 160

13. Telvin has 7 baseball trophies. How many different ways can he arrange them in a row on his bookshelf?

A. 800
B. 720
C. 4,030
D. 5,040
E. 40,320

14. Kineetha has 6 dresses, 2 pairs of shoes, and 3 purses. How many different outfits can she wear to the dance?

F. 10
G. 36
H. 32
J. 16
K. 64

288   Copyright © American Book Company

Practice Test 1

40 Minutes – 40 Questions

DIRECTIONS: Solve each problem and then choose the correct answer. Be sure to answer all the questions.

Do not linger over problems that take too much time. Solve as many as you can; then return to the others in the time you have left for this test.

You are permitted to use a calculator on this test. You may use your calculator for any problems you choose, but some of the problems may best be done without using a calculator.

Note: Unless otherwise stated, all of the following should be assumed.

1. Illustrative figures are NOT necessarily drawn to scale.

2. Geometric figures lie in a plane.

3. The word *line* indicates a straight line.

4. The word *average* indicates arithmetic mean.

---

1. Carol has saved up $78.25. She bought a shirt that cost $17.50, but is now on sale for 15% off. The pants cost $37.50, but they are also on sale now for 25% off. The shoes cost $28.50 but are not on sale, she has a coupon to get 10% off her total purchase. How much money will Carol have left over from shopping?

   A. $15.28
   B. $14.75
   C. $13.90
   D. $0.00
   E. $25.82

   PA

2. Mrs. Sanders has a class average of 82 on the Math test that she gave her students. If the highest grade on the test is a 92, what is the lowest test score on the test? (Round to nearest whole number.)

   F. 65
   G. 72
   H. 70
   J. 80
   K. Cannot be determined from the information given.

   PA

3. If $25t + 18y - 17 = 67$ and $y = 2$, what does t equal?

   A. 2.47
   B. 1.92
   C. 1.97
   D. 1.84
   E. 2.14

   PA

4. The lowest temperature on a winter morning was $-5°$F. Later that same day the temperature reached a high of $56°$F. By how many degrees Fahrenheit did the temperature increase?

   F. $56°$
   G. $60°$
   H. $61°$
   J. $65°$
   K. $53°$

   PA

Copyright © American Book Company   289

5. Karen went grocery shopping to buy meat for her annual office picnic. She bought $8\frac{5}{6}$ pounds of hamburger, 19.25 pounds of chicken, and $4\frac{2}{3}$ pounds of steak. How many pounds of meat did Karen buy?

   A. 32.67
   B. 24.64
   C. 18.27
   D. 32.75
   E. 30.23

PA

6. In scientific notation, $30,000 + 2,100,000 =$?

   F. 2.13×10^5
   G. 2.13×10^3
   H. 2.51×10^6
   J. 2.13×10^6
   K. 2.51×10^4

PA

7. Saying that $5 < \sqrt{x} < 8$ is equivalent to saying what about x ?

   A. $0 < x < 6$
   B. $12 < x < 80$
   C. $25 < x < 64$
   D. $2.5 < x < 4$
   E. $5 < x < 10$

PA

8. What value of x solves the following proportion?
   $$\frac{10}{7} = \frac{x}{14}$$
   F. 50
   G. 12
   H. 35
   J. 20
   K. 18

PA

9. On a math test, 15 students earned an A. This number is exactly 25% of the total number of students in the class. How many students are in the class?

   A. 50
   B. 80
   C. 25
   D. 60
   E. 30

PA

10. A total of 100 juniors and seniors were given a mathcmatics test. The 80 juniors attained an average score of 90 while the 20 seniors attained an average of 75. What was the average score for all 100 students who took the test?

    F. 92.3
    G. 87
    H. 50
    J. 83.6
    K. 82.5

PA

11. Five students about to purchase concert tickets for $19.45 for each ticket discover that they may purchase a block of 6 tickets for $96.00. How much would each of the 5 save if they can get a sixth person to join them and the 6 people equally divide the price of the 6-ticket block?

    A. $3.45
    B. 3.65
    C. $3.75
    D. $4.45
    E. $4.25

PA

290   Copyright © American Book Company

12. If $\left(\frac{4}{5} - \frac{3}{10}\right) + \left(\frac{2}{3} + \frac{1}{2}\right)$ is calculated and the answer is reduced to its simplest terms, what is the denominator of the resulting fraction?

F. 1
G. 2
H. 3
J. 4
K. 5

PA

13. Marsha has a bag filled with 58 marbles. 20 of them are red striped, 5 are green, 8 are blue, 7 are pink, and 18 of them are brown. What is the probability that Marsha will pick out a red striped marble and then immediately pick out a pink marble?

A. $\dfrac{16}{58}$

B. $\dfrac{70}{1653}$

C. $\dfrac{80}{2000}$

D. $\dfrac{95}{568}$

E. $\dfrac{1}{2}$

PA

14. The senior class at Richard's High School has decided to go to Disney Land for their class trip. There are 520 students in the senior class and each is only willing to pay $250.00 for the trip. The school has to pay $128,765 for the trip no matter how many students go. If 70% of the class goes, how much money will the school have to come up with?

F. $37,769
G. $38,769
H. $37,772
J. $37,765
K. $40,752

PA

15. Dr. Halifax finds the maximum heart rate (MHR) of her patients by starting with 220 beats per minute and subtracting 1 beat per minute for each year of her patients' age. Dr. Halifax recommends that her patients exercise 3 or 4 times a week for at least 20 minutes with their heart rate increasing from their resting heart rate (RHR) to their training heart rate (THR), where:

THR = RHR + 0.65(MHR – RHR)

Which of the following is closest to the THR of a 54-year-old patient whose RHR is 50 beats per minute?

A. 167
B. 192
C. 125
D. 124
E. 138

EA

16. Amanda drove her red sports car for 3 hours at a speed of x miles per hour (mph), and for 2 more hours at 78 mph. If her average speed for the entire drive was 76 mph, which of the following equations could be used to find x ?

F. $x + 78 = 2(76)$
G. $x + 2(78) = 5(76)$
H. $3x - 2(78) = 76$
J. $3x + 2(78) = 2(76)$
K. $3x + 2(78) = 5(76)$

EA

17. What is the sum of the polynomials $5a^2b + 4a^2b^2$ and $-5ab^2 + 3a^2b^2$?

A. $5a^2b - 5ab^2 + 7a^2b^2$

B. $6a^2b - 3ab + 7ab^2$

C. $5a^2b - 5ab + 7a^2b$

D. $5ab^2 + 7a^2b^2$

E. $4ab^2 - 6ab + 7a^2b$

EA

Copyright © American Book Company

18. Factor the following polynomial:

$f(x) = -x^3 + 6x^2 + 27x$

F. $x(x+3)(x-9)$
G. $x(x-3)(9-x)$
H. $x(x+3)(9-x)$
J. $x(x+9)(x-3)$
K. $x(x+3)(x-3)$

EA

19. If $5(x-3) = -25$, then $x = ?$

A. 5
B. -2
C. 2
D. -5
E. $\frac{3}{4}$

EA

20. For all nonzero r, t, and z values, $\frac{36r^3tz^5}{-4rt^3z^2} = ?$

F. $(-6)t^{-2}zr^2$
G. $(-9)t^2z^3r^2$
H. $9t^{-2}z^3r^2$
J. $(-9)t^{-2}z^3r^2$
K. $(-6)t^{-2}z^3r^2$

EA

21. Which of the following is a simplified expression equal to $\frac{25-x^2}{x-5}$ for all $x < -5$?

A. $-x-7$
B. $x + 5$
C. $x-5$
D. $-x+5$
E. $-x-5$

EA

22. What is the slope of the line with the equation $3x + 4y + 12 = 0$?

F. $\frac{1}{2}$
G. $\frac{3}{4}$
H. $-\frac{3}{4}$
J. $\frac{5}{6}$
K. $-\frac{1}{2}$

EA

23. In the figure below, if angle A is 45° and angle B is 36°, then what is the measurement of angle C?

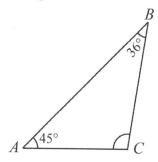

A. 54°
B. 99°
C. 101°
D. 85°
E. 97°

CG

24. The coordinates of the endpoints of AB on the real number line are 6 and 42. Point M is the midpoint of AB. What is the coordinate of M?

F. 48
G. 12
H. 24
J. 36
K. 18

CG

25. Which of the following graphs represent all, and only, the real numbers that satisfy $x + 12 \leq 26$?

A.

B.

C.

D.

E.

26. Which of the following lines in the (x, y) coordinate plane has the steepest slope?

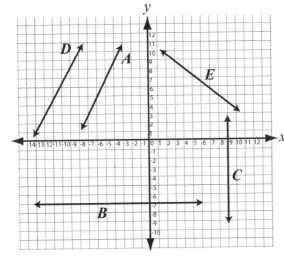

F. Line A
G. Line B
H. Line C
J. Line D
K. Line E

CG

27. Find the equation of line that passes through the points $(-1, -4)$ and $(2, -1)$.

A. $y = x - 3$
B. $y = -\frac{1}{2}x - 4$
C. $y = -\frac{5}{3}x + 3$
D. $y = -\frac{5}{3}x + \frac{7}{3}$
E. $y = 3x - \frac{7}{3}$

CG

28. Find the distance between the points $(-10, -5)$ and $(2, -7)$.

F. $\sqrt{65}$
G. $\sqrt{13}$
H. $\sqrt{37}$
J. $\sqrt{2}$
K. $2\sqrt{37}$

CG

29. Which quadrant does the point $(-3, -2)$ lie in?

A. II
B. III
C. IV
D. I
E. None of the above.

CG

30. This type of angle is greater than $90°$ but less than $180°$.

F. acute
G. right
H. straight
J. obtuse
K. scalene

PG

31. In the figure below, what is the measurement of ∠GDF?

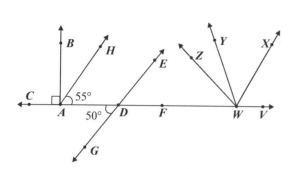

A. 110°
B. 80°
C. 130°
D. 98°
E. 180°

33. Find the area of the figure below.

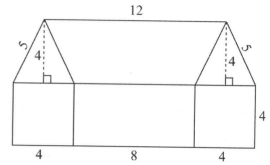

A. 48
B. 28
C. 19
D. 120
E. 46

32. In the figure below, what are the values of x and y?

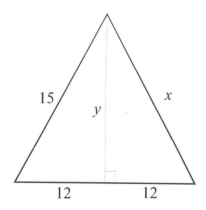

F. $x = 5, y = 12$
G. $x = 15, y = 9$
H. $x = \sqrt{3}, y = 4$
J. $x = 5, y = 9$
K. $x = 14, y = 9$

34. Find the perimeter of the figure below.

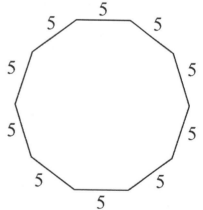

F. 55
G. 50
H. 45
J. 60
K. 65

35. Find the circumference of the circle below.

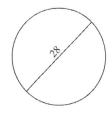

A. 87.81
B. 89.28
C. 82.97
D. 87.96
E. 78.29

PG

36. Find the area of the shaded region of the figure below.

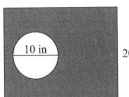

F. 486 in^2
G. 398 in^2
H. 468 in^2
J. 400 in^2
K. 422 in^2

PG

37. Find the area of the parallelogram below.

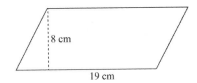

A. 188 cm^2
B. 225 cm^2
C. 180 cm^2
D. 152 cm^2
E. 160 cm^2

PG

38. What is the sum of the interior angles of a hendecagon?

F. 1,520
G. 1,620
H. 1,380
J. 1,630
K. 1,590

PG

39. What is the volume of the cylinder below?

A. $68\frac{5}{8}$ in^3
B. $62\frac{6}{7}$ in^3
C. 87 in^3
D. $81\frac{3}{8}$ in^3
E. $61\frac{6}{7}$ in^3

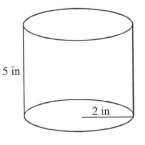

PG

40. What coordinate point will you arrive at if you start at point D and go down 3 and to the right 6?

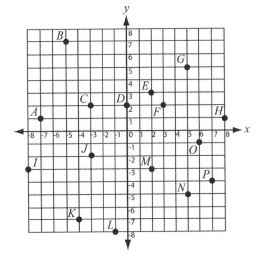

F. $(-3, -6)$
G. $(3, 6)$
H. $(-1, 6)$
J. $(0, 6)$
K. $(6, -1)$

PG

295

Evaluation Chart for the Practice Test 1

Directions: On the following chart, circle the question numbers that you answered incorrectly. Then turn to the appropriate topics (listed by chapters), read the explanations, and complete the exercises. Review the other chapters as needed. Finally, complete PLAN Mathematics Test Preparation Guide Practice Tests for further review.

		Questions	Pages
Chapter 2:	Numbers and Number Relations	12	15–24
Chapter 3:	Word Problems	4, 5, 11	25–33
Chapter 4:	Percents	1, 9, 14	34–43
Chapter 5:	Exponents and Roots	6, 7	44–58
Chapter 6:	Ratios and Proportions	8	59–65
Chapter 7:	Introduction to Graphing	24, 29	66–75
Chapter 8:	Introduction to Algebra		76–88
Chapter 9:	Equations and Inequalities	3, 19, 25	89–103
Chapter 10:	Algebra Word Problems	15, 16	104–113
Chapter 11:	Polynomials	17, 20	114–124
Chapter 12:	Factoring	18, 21	125–138
Chapter 13:	Solving Quadratic Equations		139–146
Chapter 14:	Graphing and Writing Equations and Inequalities	22, 26, 27, 28	147–171
Chapter 15:	Systems of Equations		172–187
Chapter 16:	Angles	30, 31	188–197
Chapter 17:	Triangles	23, 32	198–209
Chapter 18:	Plane Geometry	33, 34, 35, 36, 37, 38	210–228
Chapter 19:	Solid Geometry	39	229–239
Chapter 20:	Transformations	40	240–251
Chapter 21:	Statistics	2, 10	252–258
Chapter 22:	Data interpretation		259–272
Chapter 23:	Probability	13	273–288

296   Copyright © American Book Company

Practice Test 2

40 Minutes – 40 Questions

DIRECTIONS: Solve each problem and then choose the correct answer. Be sure to answer all the questions.

Do not linger over problems that take too much time. Solve as many as you can; then return to the others in the time you have left for this test.

You are permitted to use a calculator on this test. You may use your calculator for any problems you choose, but some of the problems may best be done without using a calculator.

Note: Unless otherwise stated, all of the following should be assumed.

1. Illustrative figures are NOT necessarily drawn to scale.

2. Geometric figures lie in a plane.

3. The word *line* indicates a straight line.

4. The word *average* indicates arithmetic mean.

---

1. Chris drives 228 miles and uses 10 gallons of gasoline. Mark drives 350 miles and uses 12.5 gallons of gasoline. Who has the better fuel efficiency?

   A. Chris has the better fuel efficiency at 22.8 miles per gallon.
   B. Mark has the better fuel efficiency at 28 miles per gallon.
   C. Chris has the better fuel efficiency at 28 miles per gallon.
   D. Mark has the better fuel efficiency at 22.8 miles per gallon.
   E. They both have equal fuel efficiency.

   PA

2. Which rational number is not an equivalent of $-1\frac{1}{6}$?

   F. $-\frac{7}{6}$

   G. $-1.1\overline{6}$

   H. $-\frac{14}{12}$

   J. -1.15

   K. All of the above are equivalent.

   PA

3. If you wanted to show the frequency of the number of times a quarter landed on heads or tails out of 300 flips, what type of graph should be used?

   A. circle graph
   B. line graph
   C. histogram
   D. pictograph
   E. dot plot

   PA

4. Maddy wants to join a gym. There is a $300 membership fee at the start, then a $15 a month loyalty fee. How many months can Maddy be a member at this gym if she only budgets $750 towards the gym membership?

   F. 15 months
   G. 30 months
   H. 50 months
   J. 65 months
   K. 28 months

   PA

Copyright © American Book Company

297

5. Aunt Ann uses 3 cups of oatmeal to bake 6 dozen oatmeal cookies. How many cups of oatmeal would she need to bake 15 dozen cookies?

   A. 1.2
   B. 7.5
   C. 18
   D. 30
   E. 1.8

   PA

6. If 60 students eat 24 pizzas, which proportion below may be used to find the number of pizzas required to feed 15 students?

   F. $\dfrac{60}{24} = \dfrac{15}{x}$

   G. $\dfrac{60}{24} = \dfrac{x}{15}$

   H. $\dfrac{60}{15} = \dfrac{x}{24}$

   J. $\dfrac{60}{x} = \dfrac{15}{24}$

   K. $\dfrac{x}{60} = \dfrac{1}{2}$

   PA

7. Which of the following is equivalent to the expression: $2x - 5x + 3$?

   A. $3x + 3$
   B. $x(2 - 5) + 3$
   C. $7x + 3$
   D. $2x(-5x + 3)$
   E. $x - 3$

   PA

8. Five is multiplied by a number between 8 and 15. The answer has to be:

   F. between 8 and 15
   G. between 40 and 75
   H. more than 100
   J. less than 35
   K. more than 200

   PA

9. Evaluate the following expression: $\frac{1}{4}(3x + 4)^2$ at $x = 2$.

   A. 5.5
   B. 25
   C. 30
   D. 35
   E. 52

   PA

10. Barbara has 6 lemon jellybeans, 1 licorice jellybean, 3 popcorn flavor jellybeans, and 4 root beer flavored jellybeans. What is the probability that if she takes one jellybean without looking, it is a lemon jellybean?

    F. $\dfrac{1}{2}$

    G. $\dfrac{3}{4}$

    H. $\dfrac{5}{6}$

    J. $\dfrac{1}{14}$

    K. $\dfrac{3}{7}$

    PA

11. In April, Jerry filled his car with $12\frac{1}{8}$ gallons of gas the first week, $10\frac{3}{4}$ gallons the second week, $11\frac{1}{2}$ gallons the third week, and $10\frac{1}{2}$ gallons the fourth week. How many gallons of gas did he buy in April?

    A. $32\frac{3}{4}$

    B. $43\frac{1}{2}$

    C. $30\frac{2}{3}$

    D. $44\frac{7}{8}$

    E. $41\frac{2}{5}$

    PA

298   Copyright © American Book Company

12. When Rick measures the shadow of a yard stick, it is 7 inches. At the same time, the shadow of the tree he would like to chop down is 42 inches. How tall is the tree in yards?

   F. 15
   G. 12
   H. 6
   J. 8
   K. 13

13. Using the spreadsheet below, determine how much Acme Autobody Shop spent on parts and supplies for the month of February through May.

Acme Autobody Shop Spending:		
Month	Parts/Supplies	Salaries
Jan	$2,800	$15,500
Feb	$3,400	$15,500
Mar	$4,100	$15,500
Apr	$3,250	$15,500
May	$2,890	$15,500
Jun	$3,900	$15,500
Total	$20,340	$93,000

   A. $78,400
   B. $20,950
   C. $20,340
   D. $3,900
   E. $13,640

14. Using the Venn diagram below, determine how many 9th graders chose science, math, and reading as their favorite subjects.

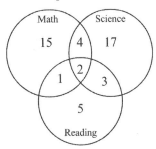
Favorite Subject of Carroll High School 9th Graders

   F. 15
   G. 17
   H. 1
   J. 4
   K. 2

15. Subtract the following polynomial.

$$(13y^3 + 2y^2 - 8y) - (2y^3 + 4y^2 - 7y)$$

   A. $y^3 + 5y$
   B. $15y^3 - 6y^2 - 15y$
   C. $11y^3 - 2y^2 - y$
   D. $2y^2 - y$
   E. $11y^3 - 2y^2$

16. Multiply the following binomial.

$$(8y - 10)(5y - 3)$$

   F. $20y^2 - 17y + 15$
   G. $40y^2 - 74y + 30$
   H. $36y^2 - 18y + 21$
   J. $18y^2 - 74y + 30$
   K. $40y^2 - 74y - 15$

17. Factor the following polynomial.

$$4m^5 + 2m^4 + 2m^3 + 6m^2$$

A. $2m^2(2m^3)$

B. $2m^3 + m^2 + m + 3$

C. $m^2 + m + 3$

D. $2m^2(2m^3 + m^2 + m + 3)$

E. $2m^2(m^3 + m^2 + m + 3)$

EA

18. At Summer High School, the probability of a student taking statistics this semester is 32%. The probability of a student taking statistics and chemistry at the same time is 12%. What is the probability that Don will take chemistry this semester given that he is taking statistics?

F. 0%

G. 25%

H. 34.5%

J. 37.5%

K. 40.2%

EA

19. Factor: $x^3 - 9x^2 + 27x - 27$.

A. $(x - 9)^3$

B. $(x - 3)^3$

C. $(x - 7)^3$

D. $(x - 8)^3$

E. $(x - 5)^3$

EA

20. Solve: $t^2 - 11t + 30 = 0$.

F. $\{3, 8\}$

G. $\{2, 5\}$

H. $\{5, 8\}$

J. $\{10, 17\}$

K. $\{5, 6\}$

EA

21. If $x < 2$, simplify: $|x - 2| - 4| - 6|$

A. $x + 22$

B. $-x + 22$

C. $-x - 22$

D. $x - 2$

E. $-x - 2$

EA

22. Sharice is a waitress at a local restaurant. She makes an hourly wage of $9.50, plus she receives tips. On Monday she works 8 hours and receives tip money, t. Write an equation showing what Sharice makes on Monday, y.

F. $y = \$9.50 \times t + 8$

G. $y = t \times 8 + \$9.50$

H. $y = \$9.50 \times 8 + t$

J. $y = \$9.50 \times 8t + 8$

K. $y = \$9.50 \times 8 + \$9.50t$

EA

23. In which quadrant does the point $(2, 3)$ lie?

A. I

B. II

C. III

D. IV

E. cannot be determined

CG

Copyright © American Book Company

24. Which of the following is not a solution of $3x = 5y - 1$?

   F. $(3, 2)$
   G. $(7, 4)$
   H. $\left(-\frac{1}{3}, 0\right)$
   J. $(-2, -1)$
   K. $\left(0, \frac{1}{5}\right)$

25. What is the x-intercept of the line $y = 5x - 6$?

   A. $(3, 0)$
   B. $\left(\frac{6}{5}, 0\right)$
   C. $(-3, 0)$
   D. $\left(\frac{1}{2}, 0\right)$
   E. $\left(-\frac{5}{6}, 0\right)$

26. The coordinates for the endpoints of a line segment are $(-5, 13)$ and $(-9, 21)$. Find the coordinates for the midpoint of the line segment.

   F. $(-7, 17)$
   G. $(-4, 4)$
   H. $(-2, 4)$
   J. $(4, 17)$
   K. $(3, 12)$

27. Which is the graph of $x - 3y = 6$?

A.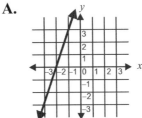

B.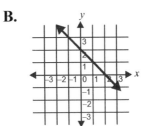

C.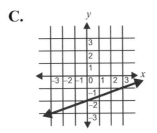

D.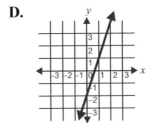

E.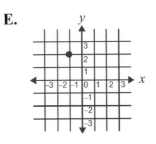

28. What is the relationship between ∠1 and ∠3?

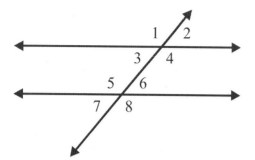

F. They are corresponding angles.
G. They are vertical angles.
H. They are alternate interior angles.
J. They are supplementary angles.
K. They are alternate exterior angles.

CG

29. Three points that lie on the same plane are always

A. coplanar.
B. collinear.
C. collinear and coplanar.
D. congruent.
E. congruent and collinear.

CG

30. Find the missing side from the following similar triangles.

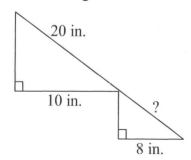

F. 8
G. 16
H. 24
J. 32
K. 46

PG

31. Angle P and angle Q are complementary angles. The measure of angle P is $3x + 5$ and the measure of angle Q is $4x - 20$. What is the measure of angle P in degrees?

A. 40°
B. 50°
C. 80°
D. 85°
E. 60°

PG

32. In the diagram below, which line segment represent the radius of the circle?

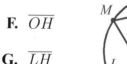

F. $\overline{OH}$
G. $\overline{LH}$
H. $\overline{MN}$
J. $\overline{JP}$
K. $\overline{MJ}$

PG

33. Which is true about a rectangle?

A. All four sides of a rectangle are equal.
B. The opposite sides in a rectangle are parallel to each other and are equal.
C. All of the sides are of different lengths.
D. The opposite sides in a rectangle are not parallel to each other, but they are equal.
E. All rectangles are squares.

PG

34. What is the area of the triangle below?

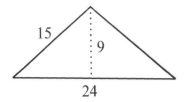

F. 54 square units
G. 67.5 square units
H. 108 square units
J. 216 square units
K. 97.6 square units

PG

35. The two triangles below are congruent by

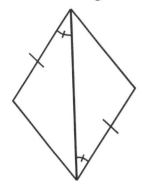

A. AAA
B. SSS
C. ASA
D. SAS
E. AAS

PG

36. Find the perimeter of the trapezoid.

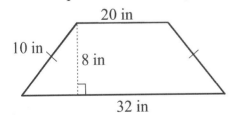

F. 70 in
G. 208 in
H. 105 in
J. 860 in
K. 72 in

PG

37. What is the $m\overline{AB}$?

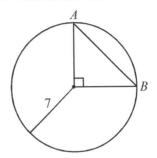

A. 7
B. $7\sqrt{2}$
C. $7\sqrt{3}$
D. $\sqrt{2}$
E. $7\sqrt{5}$

PG

38. What is the relationship between these two lines?

$y = -\frac{5}{3}x + 7$

$y = \frac{3}{5}x + \frac{3}{5}$

F. parallel
G. perpendicular
H. coincident (same time)
J. intersecting, but not perpendicular
K. cannot be determined by the information given

PG

39. What is the value of x? Lines l and m are parallel to each other.

A. 75°
B. 95°
C. 105°
D. 150°
E. 175°

PG

40.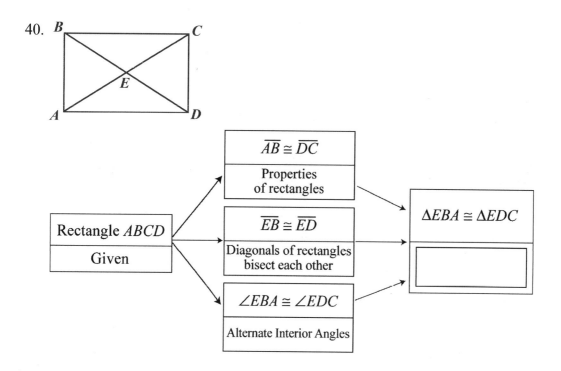

This flowchart proof shows $\triangle EBA \cong \triangle EDC$ by

F. substitution property of equality
G. side-angle-side
H. symmetric property
J. CPCTC
K. side-side-angle

Evaluation Chart for Practice Test 2

Directions: On the following chart, circle the question numbers that you answered incorrectly. Then turn to the appropriate topics (listed by chapters), read the explanations, and complete the exercises. Review the other chapters as needed. Finally, complete PLAN Mathematics Test Preparation Guide Practice Tests for further review.

		Questions	Pages
Chapter 2:	Numbers and Number Relations	2, 4, 8	15–24
Chapter 3:	Word Problems	1, 11	25–33
Chapter 4:	Percents		34–43
Chapter 5:	Exponents and Roots		44–58
Chapter 6:	Ratios and Proportions	5, 6, 12	59–65
Chapter 7:	Introduction to Graphing	23	66–75
Chapter 8:	Introduction to Algebra	9, 22	76–88
Chapter 9:	Equations and Inequalities	7, 21	89–103
Chapter 10:	Algebra Word Problems		104–113
Chapter 11:	Polynomials	15, 16	114–124
Chapter 12:	Factoring	17, 19	125–138
Chapter 13:	Solving Quadratic Equations	20	139–146
Chapter 14:	Graphing and Writing Equations and Inequalities	24, 25, 26, 27	147–171
Chapter 15:	Systems of Equations	38	172–187
Chapter 16:	Angles	28, 31, 39	188–197
Chapter 17:	Triangles	30, 35, 37, 40	198–209
Chapter 18:	Plane Geometry	29, 32, 33, 34, 36	210–228
Chapter 19:	Solid Geometry		229–239
Chapter 20:	Transformations		240–251
Chapter 21:	Statistics		252–258
Chapter 22:	Data interpretation	3, 13, 14	259–272
Chapter 23:	Probability	10, 18	273–288

Copyright © American Book Company

Index

Acknowledgements, ii
Addition
  of polynomials, 116
Algebra
  multi-step problems, 96
  two step problems, 89
    with fractions, 90
  vocabulary, 76
  word problems, 78, 104
    age problems, 108
    setting up, 81
Alternate Exterior Angles, 193
Alternate Interior Angles, 193
Angles, 188
  adjacent, 190
  alternate exterior, 193
  alternate interior, 193
  complementary, 192
  corresponding, 193
  linear pair, 192
  sum of polygon interior, 194
  supplementary, 192
  vertical, 191
Area
  circle, 218
  squares and rectangles, 214
  triangles, 215
  two-step problems, 219
Associative Property
  of addition, 84
  of multiplication, 84

Bar Graphs, 262
Base, 76
Binomials, 114, 130
  multiplication
    FOIL method, 120

Cartesian Plane, 147

Cartesian plane
  coordinate quadrants, 71
Circle Graphs, 265
Circumference, Circle, 217
Coefficient, 76
  leading, 76
Collinear Points, 152
Combination
$$_nC_r = \frac{n!}{(n-r)!r!}, 283$$
Combining like terms, 92
Commutative Property
  of addition, 84
  of multiplication, 84
Complementary Angles, 192
Constant, 76
Corresponding Angles, 193
Corresponding Sides
  triangles, 202

Decimals
  changing to percents
    changing percents to decimals, 34
Degree, 76
Dependent Events, 278
Diagnostic Test, 1
Diameter, Circle, 217
Difference of Two Squares, 141
Dilation, 247
Discount, 40
  finding the sale price, 41
Distance Formula
$$d = \sqrt{(y_2 - y_1)^2 + (x_2 - x_1)^2}, 150$$
Distributive Property, 84

Equations
  finding using two points or a point and
    slope, 161
  linear, 147
  linear systems

306

solving by adding or subtracting, 180
    solving by substitution, 178
  of perpendicular lines, 173
  solving with like terms, 92
Estimated Solutions, 29
Estimating, 28
  rational numbers, 28
Evaluation Chart, 8, 296, 305
Exact Solutions, 29
Exponents, 44, 114
  dividing, 49
  multiplying exponents raised to an exponent, 46
  multiplying polynomials, 117
  multiplying with negative exponents, 48
  multiplying with same base, 46
  negative exponents, 48
  of polynomials, 115
  simpifying binomial expressions, 121
  when subtracting polynomials, 116

Factor, 16
Factoring
  difference of two squares, 135
  of polynomials, 125
  quadratic equations, 140
  trinomials, 130
FOIL Method
  for multiplying binomials, *see* Binomials
FOIL method, 130
Fractions
  changing percents to fractions
    changing fractions to percents, 35
  ordering, 22
Frequency Table, 259

Geometry
  word problems, 107
Graphing
  a line knowing a point and slope, 159
  fractional values, 66
  horizontal and vertical lines, 149
  inequalities, 163
  linear equations, 147

  on a number line, 66
Graphs
  bar, 262
  circle, 265
  line, 263
  pictographs, 266
  pie, 265

Histogram, 260
Hypotenuse, 204

Identical Lines, 175
Identity Property
  of multiplication, 84
  of zero, 84
Independent Events, 278
Inequalities
  compound, 98
  graphing, 98
  multi-step, 99
  solution sets, 98
  word problems, 111
Inequality
  definition, 76
Integers, 117
  word problems, 25
Intercepts of a Line, 154
Intersecting Lines, 175, 177
Inverse Property
  of addition, 84
  of multiplication, 84
Irrational Numbers, 15

Key (Legend), 266

Least Common Multiple, 19
Legend (Key), 266
Line, 210
Line Graphs, 263
Line of Reflection, 240
Line Segment, 210
Linear Equation, 147, 157
Linear Pair, 192
Lines

307

identical, 175
intersecting, 175, 177
parallel, 172, 175

Mean, 253
Measures of Central Tendency, 255
Median, 254
Midpoint of a Line Segment
$M = \left(\frac{x_1+x_2}{2}, \frac{y_1+y_2}{2}\right)$, 151
Mode, 254
Monomials, 114
adding and subtracting, 114
multiplying, 117
multiplying by polynomials, 118
Multi-Step Algebra Problems, 96

Number line, 66
vertical, 68
Number Lines, 69
graphing rational numbers, 69

Order of Operations, 50
Please Excuse My Dear Aunt Sally, 50
Ordered Pair, 147
Ordered pairs, 72
Origin, 71

Parallel lines, 172, 175
Parentheses
removing, 95
Parentheses, Removing and Simplifying
polynomials, 119
Patterns
number, 101
Percents, 34
changing percents to decimals
changing decimals to percents, 34
changing percents to fractions
changing fractions to percents, 35
decrease or increase, 38
finding percent of the total, 37
finding the amount of discount, 40
finding the discounted sales price, 40
Perfect Squares, 135, 143

Perimeter, 107, 213
Permutation
$_nP_r = \dfrac{n!}{(n-r)!}$, 280
Perpendicular Lines
equations of, 173
π, 217
Pictographs, 266
Point-Slope Form of an Equation
$y - y_1 = m(x - x_1)$, 161
Points
collinear, 210
coplanar, 210
Polygons, 213
parallelogram, 212
quadrilateral, 212
rectangle, 212
rhombus, 212
square, 212
sum of interior angles, 194
trapezoid, 212
Polynomial(s), 114, 125
adding, 115
factoring, 125
greatest common factor, 125–127
multiplying by monomials, 118
subtracting, 116
Practice Test 1, 289
Practice Test 2, 297
Preface, xi
Prime Factorization, 17
Prime Numbers, 17
Probability, 273
Product, 125
Proportions, 60
Pythagorean Theorem, 204
applications, 206
Pythagorean Triples, 204

Quadratic equation
$ax^2 + bx + c = 0$, 144
Quadratic Equations, 139
Quadratic formula

308

$$\frac{-b \pm \sqrt{b^2 - 4ac}}{2a}, 144$$

radical, 54
Radius, Circle, 217
Range, 252
Rational Numbers, 15
Ratios, 59
Ray, 210
Real Numbers, 15
Reasonable Solutions, 30
Rectangle, 107
Reflection, 240
  line of, 240
Reflexive Property of Equality, 84
Right Triangle, 204
Rotation, 245

Scale Factor, 247
Scientific Notation, 52
  for large numbers, 52
  for small numbers, 53
Sentence, 76
Sequences, 101
Similar Figures, 221
Similar Triangles, 202
Slope, 157, 175
$$m = \frac{y_2 - y_1}{x_2 - x_1}, 155$$
Slope Intercept Form of a Line
  $y = mx + b$, 157, 175
Solid Geometry Word Problems, 235
Square Roots
  estimating, 54
Square roots, 54
Statistics, 252
Substitution
  numbers for variables, 77
Subtraction
  of polynomials, 116
Subtrahend, 116
Supplementary Angles, 192
Symmetric Property of Equality, 84

Table of Contents, x

Tables
  data, 261
Tally Chart, 259
Term, 76
Transitive Property of Equality, 84
Translation, 243
Transversal, 193
Tree Diagram, 276
Triangle, 107
  isosceles, 107
Triangle Inequality Theorem, 201
Triangles
  corresponding sides, 202
  exterior angles, 200
  proportional sides, 202
  similar, 202
  sum of interior angles, 199
Trinomials, 114
  factoring, 130
Two Step Algebra Problems, 89
  with fractions, 90
Two-Step Area Problems, 219
Two-Step Volume Problems, 233

Variability, 252
Variable, 76, 93, 114, 115, 117
Venn Diagram, 267
Volume
  rectangular prisms, 229
  rectangular solids, 229
  spheres, cones, cylinders, and pyramids, 231
  two-step problems, 233

Word Problems
  algebra, 78
    setting up, 81
  changing to algebraic equations, 82
  fractions, 26
  geometry, 107
  ratios and proportions, 61
  solid geometry, 235
  solving with systems of equations, 182

y-intercept, 175